The behavioural environment

Geography has a central role 'in the ultimate purpose of all intellectual endeavour – the heightening of human consciousness'. The behavioural environment model, as developed by William Kirk, emphasizes the importance of perception in human geography, the significance of subjective experience, and the potential of man as an active agent in the environment.

This collection of essays examines the concept of the behavioural environment. Drawing on Kirk's work, the book reflects on the original formulation of the theory, applies its cardinal principals in different contexts, and re-evaluates its cognitive claims in the light of recent debates in philosophy and social theory. The book presents new empirical studies and fresh theoretical perspectives and argues that the behavioural approach is vital to a consistent understanding of individual rights and ecological respect.

The Behavioural Environment is not an unqualified acclamation of Kirk's particular approach. Crucial to the book are criticisms, extensions and modifications of the concept. The contributors address the central theme from a variety of philosophical stances, ranging from idealism and phenomenology to Marxism and critical theory, and in a variety of empirical arenas, including ethnic patterns of awareness in Jerusalem and the social management of Victorian lunacy. The coherence of the collection lies in the idea of the behavioural environment itself as a focal point for the diverse expressions of a reconstituted human geography.

The editors
Frederick W. Boal is a reader in Geography and Director of the Centre of Canadian Studies at the Queen's University of Belfast. He is co-editor of *Integration and Division: Geographical Perspectives on the Northern Ireland Problem*.

David N. Livingstone is senior research officer in Geography at the Queen's University of Belfast. He is the author of *Nathaniel Southgate Shaler and the Culture of American Science* and *Darwin's Forgotten Defenders*, both 1987.

William Kirk, OBE
1921–1987

THE BEHAVIOURAL ENVIRONMENT

Essays in Reflection, Application, and Re-evaluation

Edited by
Frederick W. Boal
and David N. Livingstone

Routledge
Taylor & Francis Group

LONDON AND NEW YORK

First published 1989 by Routledge
2 Park Square, Milton Park, Abingdon, Oxfordshire OX14 4RN
711 Third Avenue, New York, NY 10017

First issued in paperback 2015

Routledge is an imprint of the Taylor and Francis Group, an informa business

© 1989 Frederick W. Boal and David N. Livingstone

Typeset by J&L Composition Ltd, Filey, North Yorkshire

British Library Cataloguing in Publication Data

The behavioural environment: essays in reflection,
  application and re-evaluation.
  1. Geography. Behavioural aspects
  I. Boal, F. W. (Frederick Wilgar, *1934–*)   II.
  Livingstone, David N. *1953–*
  910'.01'9
  ISBN 0–415–00454–3

Library of Congress Cataloging-in-Publication Data

The Behavioural environment.
  In memory of William Kirk.
  Bibliography: p.
  Includes index.
  1. Geographical perception.   2. Anthropo-geography.
  3. Kirk, W. (William)   I. Boal, Frederick Wilgar.
  II. Livingstone, David N., 1953–   .   III. Kirk, W.
  (William)
  G71.5.B44   1989   304.2   89–10511
  ISBN 0–415–00454–3

ISBN 13: 978-1-138-88128-0 (pbk)
ISBN 13: 978-0-415-00454-1 (hbk)

To the memory of
William Kirk

Contents

Contents

Figures

Figures

Tables

Contributors

T. J. Anderson: University of Ulster at Jordanstown
Frederick W. Boal: The Queen's University of Belfast
Harold Brookfield: Australian National University
Anne Buttimer: University of Lund
John A. Campbell: The Queen's University of Belfast
John R. Gold: Oxford Polytechnic
Margaret M. Gold: Ealing College of Higher Education
Brian Goodey: Oxford Polytechnic
Leonard Guelke: University of Waterloo
R. J. Johnston: University of Sheffield
The late *William Kirk:* The Queen's University of Belfast
David N. Livingstone: The Queen's University of Belfast
Chris Philo: St David's University College, Lampeter
Douglas Pocock: University of Durham
Edward Relph: University of Toronto
Michael Romann: Tel Aviv University
O. H. K. Spate: Australian National University
Yi-Fu Tuan: University of Wisconsin-Madison
Wreford Watson: University of Edinburgh

Preface

Geographers, as William Kirk often reminded us, have their own particular role to play 'in the ultimate purpose of all intellectual endeavour – the heightening of human consciousness'. Indeed, it was by focusing on human consciousness itself that Kirk made his own distinctive contribution to geography's intellectual enterprise. In introducing geographers to what he called 'the behavioural environment', he brought perception to the centre stage of geographical investigation by arguing the case for the prime importance of reconstructing the perceived worlds of the cultural groups and individual decision-makers that they studied. But more than that, by sensitizing geographers to the all-pervasiveness of perception in human affairs, it became difficult to escape the implication that they themselves were subject to the influence of their *own* behavioural environments. There are, then, two distinct but related cognitive themes that reverberate within the overall behavioural environment model – namely, the behavioural environment as *itself* crucial to human geographical cognition and, second, the philosophical implications of *knowledge* of the behavioural environment for the study of geography.

Consequently this volume draws together a series of essays which take Kirk's original schema as their point of departure, some reflecting on its origin and evolution, others applying its cardinal principles within a variety of contexts, still others re-evaluating its central cognitive claims in the light of contemporary debates in philosophy and social theory. Drawing substantially on the tradition of non-quantitative human geography, authors suggest ways in which the notion of the behavioural environment may enable geographers to take more seriously into account the significance of subjective experience and of people as active agents. The philosophical assumptions from which this central problem is addressed are diverse; but the value of retaining or reforming the behavioural environment model constitutes the volume's leitmotif. And just as the philosophical styles are diverse, so too are the empirical arenas within which the schema is deployed. Accordingly, while theorists run the gamut of opinion from idealism and phenomenology to Marxism and critical theory, the sites of empirical engagement range from housing in Belfast, ethnic patterns of awareness in Jerusalem, and perception in the non-Western world to the ideology of British architecture, small-town images, and the social management of Victorian lunacy.

The Behavioural Environment is assuredly not an unqualified acclamation of Kirk's particular approach. Criticisms, extensions, and modifications of the concept are as crucial to this volume as are reflections on its history and reviews of its current applications. The coherence of this collection, therefore, lies not in any unified method or specific philosophical stance, but in the idea of the behavioural environment itself as a focal point for the diverse expressions of a reconstituted human geography.

The idea of gathering together a series of essays on the behavioural environment was originally conceived by the editors during the last year of William Kirk's tenure of the Chair of Geography at the Queen's University of Belfast. He readily agreed to the republication of his original 1952 paper and was eagerly looking forward to perusing the evaluations of his schema by colleagues of long standing as well as by those of a newer generation. Sadly, Bill did not live to see the completion of the project.

The esteem in which Bill Kirk was held within the international geographical community is reflected in the warmth and alacrity with which all our authors responded to the invitation to contribute to this volume. Without their efforts this broad-ranging collection of perspectives on the behavioural environment theme would obviously have been impossible. Neither, however, could a set of essays such as this have been produced without the generous help of many friends. Particularly appreciative words of acknowledgement must go to Jenitha Orr for her painstaking scrutiny of the typescript at every stage of its transformation into a unified text and her unfailing watchfulness in ensuring that the editors followed the right stylistic rules. Our thanks also go to Suzanne Geddis and Velma Atcheson for transferring each chapter to disc as it arrived, and to Gill Alexander and Maura Pringle for making their cartographic skills available to us. We are grateful to the Queen's University of Belfast which kindly provided a grant from its Publications Fund towards the cost of producing this volume. Finally we thank Mrs Laura Kirk for providing the photograph used as the frontispiece and express our hope that she will find in these essays a fitting tribute to her husband's scholarly career.

Fred Boal
David Livingstone
The Queen's University of Belfast

Foreword

O. H. K. Spate

For one of an older generation of geographers, working before the flood of quantification, regional analysis, and the like seemed to sweep away the old landmarks, it is indeed flattering to be invited to contribute a foreword to a volume which expresses in a variety of ways the humanistic view of geography towards which I had always been groping. Although I can still recall the excitement with which I read Kirk's papers on the behavioural environment when they first appeared, for two decades my interests have lain in a different discipline, history, and this is, as it were, a recall to an old duty, which is also a pleasure.

William Kirk's essay on 'Historical geography and the behavioural environment' appeared some three and a half decades ago; perhaps both the place and time of its publication were unfavourable to a recognition of its value. As for place, its appearance in an Indian volume combining a jubilee souvenir and a memorial tribute, to one unknown outside India, deprived it of the impact it might have had if it had been published in, say, the *Transactions of the Institute of British Geographers*; as for time, British geography was hardly emerging from the remarkable theoretical apathy, exceeding the norm of English empiricism, which marked, but hardly distinguished, the period between the wars.

The Institute used to devote its Sunday evening after-dinner session at the annual conference to a discussion of methodology, a token recognition that there was such a thing. We younger people were wrapped up in what now seem naïve arguments about possibilism and determinism; as for the Old Guard, its attitude was perfectly summed up by the remark of one of them to me after one such discussion: 'All this theoretical stuff is very interesting, but then one would have to think, and one is tired.' Theory was something that odd Germans might be interested in, and should be left to them – a stultifying view, or lack of view.

Once started, however, the war-horse of theory bolted into the fray, and those of us who were not obscurantists but had no training in the new quantitative techniques were left somewhat bewildered. It is ironic, but usual in the history of revolutions, that precursors who had argued for the necessity of change if the discipline were to have any validity in the discourse of ideas, of breaking out of the deadly morass of mere empiricism, should find that the change came in unexpected and sometimes rather awesome forms. We had called for some philosophy, and were swamped by epistemology. The revolution, again as is the habit of revolutions, developed a dynamic momentum of its own, and those who could not cope, like myself, had little option but to retreat while we could still do so with some dignity.

In all this I do not for one moment wish to deny the great achievement of quantification, systems analysis, and the rest, which have given a new face to geography, now more normative than of old – even if some exciting discoveries seem a bit trite, and 'simulation models' for example sometimes bear an uncanny resemblance to what we used to call 'hypothetical diagrams'. The heady new wine broke a lot of old bottles, but there were other vintages in reserve, awaiting maturation. There was always faith that older humanist values, where they had real value, would re-emerge and that there would be not merely coexistence (even that seemed threatened by the more ardent spirits for a time) but collaboration and some degree of synthesis of the old and the new.

New lights came from philosophy, in an interest in epistemology and causation; from psychology, in theories of cognition and conceptualization: both essential, fundamental, to behavioural geography. It had always been recognized, but vaguely, that 'the facts are as apprehended', and apprehension varies with varying environments in space and time. After all, Humboldt himself had written, in Part II of *Cosmos*, that wide-ranging and perceptive essay in topophilia, 'The differences of feeling excited by the contemplation of nature at different epochs, and amongst different races of men'. There had also been Michelet's marvellously evocative *Tableau de la France*, still dimly remembered from my schooldays. Now we have from America the work of the Sinopsychogeographer Yi-Fu Tuan and the memorial volume to J. K. Wright edited by David Lowenthal and Marilyn Bowden under the apt and evocative title *Geographies of the Mind*. The role of cognition has been at once extended, deepened, and sharpened. A new revolution has arisen alongside the quantitative one, and they are not without points of contact.

The environmentalism of the first couple of decades of this century, by Ellen Churchill Semple out of Ratzel, was disdained by mainstream geographers of the thirties, at least in Anglo-Saxon countries. They pinned their faith, when they had any, in a wishy-washy possibilism, under the influence of Vidal de la Blache and Lucien Febvre, whose misunderstood writings were dissected by us undergraduates of those days with meticulousness usually reserved for Holy or Marxist Writ. There was justice in this disdain of Ratzelian 'influences of the geographical environment', which all too easily slipped into the crudest determinism. However, the Old Guard had nothing to put in its place, at least on the human side of the subject, excepting a few things such as J. F. Unstead's attempt to work out a hierarchy of regions. The good work that was done, and there was a good deal of it, was essentially in a descriptive mode, rarely addressing itself to real issues of the world's life. C. A. Fisher had even to plead, and after the war at that, for some economics in economic geography, and met with rather a shocked response ('nous avons changé tout cela'). However, environmentalism, then practically a term of contempt, is now a potent word, and the new environmentalism is far more supple and subtle than the old.

In the new environment of thought, the seed which Kirk sowed so long ago, and which seemed to have fallen by the wayside, has found its soil, taken root, and, as this volume amply displays, is flourishing. The human role in the shaping of geography is no longer seen as simply the mechanical activities which have changed the aspect of vast areas of the earth's surface, often for the worse, but in a more intimate way, if also a more indirect one. The converse holds: environmental influence can no longer be seen as a direct object-to-object impact (the source of some of the cruder extravagances of the old determinism) but as filtered and refracted not only through economic and political activities, but also through the modes of human consciousness and perception, of archetypal symbolism and of cultural tradition. Towards this some of us were already groping back in the thirties, but it was usually incidental to regional studies, and there was no principle, no clear programme for our incipient ideas.

It is obvious that the concept of the behavioural environment is central to the development of a consistent approach, to giving coherence to what were only suggestive notions, half-starts. Along with the respect for peoples in their own right and the insistence on ecological understanding, the behavioural approach has a significant part to play in the struggle to maintain individuality and difference – in short, human dignity – and to restore health to our ravaged earth. Kirk's work was, as it were, the catalytic crystal in the saturated solution. This volume gives ample evidence of the range and power of the concept, and of its high degree of applicability to some of the pressing problems of our time.

I am sure that *The Behavioural Environment* will stand as a striking exemplar of the value of the concept, which gives the book a unity not usual in edited collections, an elegant as well as a valuable tool for our understanding of the world and our place in it.

Part one

The behavioural environment

Chapter one

The behavioural environment: worlds of meaning in a world of facts

Frederick W. Boal and David N. Livingstone

Historians and philosophers ordinarily interrogate the past in rather different ways. For historians, we might say, the objective is to find out about what happened in the past; the philosopher's task, by contrast, is to make the best use of past discourses to shed light on contemporary problems. In the history and philosophy of geography these different aims have too often been confused. And the result is that many histories of the discipline are presentist in taste and Whiggish in tone; that is to say, they depict geography's history as a story leading inexorably to present-day orthodoxy, suppressing themes that lack contemporary respectability and ignoring those blind alleys that putatively deviate from the 'proper' course of historical development. Such histories amount to little more than propaganda for some particular orthodoxy and it is no surprise that there are just about as many presentist histories of geography as there are definitions of the subject.

To appreciate the pitfalls of history written backwards, written indeed to legitimate partisanship, does not however invalidate the philosophical task of seeking answers to contemporary problems through probing the riches of a discipline's intellectual heritage. Just as philosophers still find it profitable to reread Aristotle or Descartes in their perennial attempt to figure out how things are, so geographers should not apologize for making the best use of their own traditions for similar purposes. To sacralize modernity (the 'only-what's-published-today-counts' syndrome) is no less ensnaring than to mutilate history. To be sure, scrutinizing the past for philosophical ends can too easily result in attributing false motives to earlier writers. Distortion, or indeed manipulation, of this sort is clearly to be avoided. But that danger should never blind us to the value of looking to our intellectual dowry for the resources to grapple with the present and the future. By fitting on the garments of the past, trimming here, enlarging there, we may find the where-withal to make significant contributions to contemporary needs.

It is in the spirit of philosophical scrutiny, therefore, that we approach the notion of the behavioural environment in modern geography: for embedded in that concept are a number of crucial philosophical issues that confront contemporary human geography, questions like the relationship between society and environment, mind and world, realism and anti-realism, culture and nature, structure and agency. Certainly, the philosophical concerns of today's geography are far from identical with those of the 1950s and 1960s when the idea of a 'behavioural environment' first began to receive an airing in the discipline. However, the articulation of the notion of the behavioural environment, and the sorts of problem it was designed to meet, are related, albeit mostly by family resemblance, to key conceptual questions on geography's present-day philosophical horizon.

Nature, humanity, and geographical explanation

The very juxtaposition of the components in the label 'human geography' taken in literal simplicity as humankind and earth-description, conceals a profound quandary that lies at the heart of geographical enquiry – namely, how we are to conceive of the relationship between the human and natural worlds. For R. J. Chorley, writing in 1973, the development of geography remained 'bound up with one of the basic philosophical problems which preoccupied the medieval world, namely to what extent is it proper to regard Man as a part of Nature or as standing apart from it?'[1] In the West, very broadly speaking, two different answers have been offered – two perspectives or pictures about humanity and the world. The first has been available since ancient times; the second is a more recent creation, emerging clearly during the Enlightenment. For convenience we might label these 'Naturalism' and 'Humanism' respectively, and it is no surprise that these are chief among the explanatory accounts that are still with us. Accordingly a word about each is in order.[2]

In its essentials naturalism goes back to figures like Epicurus and Democritus and finds purposeful expression in Lucretius's *De Rerum*. The basic idea is that the best way to understand ourselves and what seems to be distinctive about us is ultimately in terms of our community with non-human nature. We are part of nature, and no more; and everything about us can be reduced to material constituents. To be sure, the ways in which this outlook manifests itself are complex and diverse. In our time the perspective is broadly evolutionary: the human condition is entirely explicable in terms of random genetic mutation. Altruism, for instance, is to be

understood as nothing but a strategy for safeguarding the geno-
type; loyalty, it would seem, is no less genetically coded for than
ear lobes. In an earlier day, and still in some quarters today,
mechanistic metaphors were more commonly resorted to than the
organic analogy, and were often construed deterministically. Here,
human beings are conceived of as just sophisticated machines,
their brains cybernetic, their behaviour mechanistic. As for method,
the implication is usually taken to be that the way in which the
world of nature is to be studied is precisely the same for humankind: there is one scientific method universally applicable, whether
to molecules or microbes, culture or class.

The 'humanist' alternative is quite different. It tends to place
emphasis on the human subject as, in some deep and important
way, responsible for the structure and nature of the world. This
view goes under a number of different labels: 'idealism' is one,
'Enlightenment humanism' another, 'anti-realism' a different,
and perhaps better designation. By and large we can trace the
roots of this vision to the writings of Immanuel Kant. According to
him the external world owes its very structure to what we might
call the noetic activity of persons. The idea is that the human mind
imposes its own structures on the world; space and time, causality,
number, and so on – these are not to be found in the world as such,
but are conferred on it by mind. Of course it might be difficult to
think of any sort of thing existing that did not have at least some of
these very properties, and if so then in a real sense there would be
no entities in the world if it were not for the structuring activity of
persons like us. At any rate if there truly is something out there it
is a numinous shadowy world bearing no relationship to how we
conceive of it. Since Kant, vast stretches of modern philosophy
have remained staunchly anti-realist in impulse: much European
continental thought continues the tradition, the Wittgensteinians
reveal it in their version of linguistic anti-realism, and in the
United States it has been vigorously restated by pragmatists like
Richard Rorty. The idea is simply that by virtue of our activities –
linguistic, cognitive, theoretical, or whatever – *we* are the creators
of the world.

Needless to say, these two broad strands of thought are far from
homogeneous in expression. Moreover, while they are in fact
largely antithetical they are to be found in every conceivable sort
of combination. This is due, at least in part, to some of the
difficulties that each of them faces independently. For some the
problem with naturalism is that it vastly underestimates the role of
human beings in the scheme of things, whereas humanism vastly
overestimates it. Certainly there are many other philosophical

5

queries that each need to address, but for our purposes these two problems seem relevant, particularly when they manifest themselves in the form of reductionism on the one hand, and relativism on the other. The former is essentially a naturalistic strategy by which observed behaviour on any particular level can be reduced to, or explained by, the processes of a more basic layer; hence there are attempts to reduce the social to the biological, the biological to the chemical, the chemical to the atomic. Moreover, within any particular layer reductionism is also at work: at the sociological level, for example, human behaviour has often been accounted for in terms of sociological programming, behaviouristic stimulus–response mechanisms, or economic conditions.

In geography the self-same spirit has found expression. The resort to environmental determinism in its various colours to account for human spatial behaviour is a notable instance: here history is reduced to nature. The more fundamentalist varieties of Marxist geography which reduce social life to economic base are similarly inspired. Indeed, when these Marxists account for the flourishing of environmental determinism as nothing but the expression of ideological interests they are substituting one form of reductionism for another. And there are other manifestations too: the explanation of aggression in terms of biological territorialism, the development of gravity models within the context of a geographically honed social physics, or the creation of 'rational economic man'. All these, to one degree or another, display the naturalistic impulse towards reductionism. The suppression of human agency that these fatalistic programmes engage has, of course, been the subject of numerous critiques that need not be rehearsed here. We need only recall that all these ideas have a *social* history embodying particular ideological allegiances and that they have invariably resorted to 'nature' or to some 'natural law' as the legitimating source of their socio-political agenda.[3]

If the naturalistic ethos is typically implicated in reductionist modes of thought and action, then the humanist alternative frequently espouses some form of relativism. The argument here regularly runs along lines something like this. If the world is somehow structured by the cognizing of human beings, then it may be that different humans construct it in different ways. The lived world, or *Lebenswelt* as it is often styled, of one individual or group may bear no relationship to that of another – and *a fortiori* to the world itself. Indeed, the moral is often drawn that even if these worlds – the internal and external, so to speak – did match up, we would not be able to know that they did. The very idea of comparing one system of beliefs with unconceptualized reality just simply makes no sense. The underlying conviction at work here is that the whole idea of truth, of things being some way, is an illusion. Instead, truth is taken to be relative to knowers, relative because the languages, for example, that human beings use constitute the way the world is. Language-users are, therefore, world-makers. And herein lies the fundamental reason why the anti-realist notion of truth entails the thesis that the world is actually constituted by humans. If there is no correspondence between our words and things in the world – say, ducks or drumlins – these entities would not exist if it were not for us and our conceptualizing activities. To be sure there might be some undifferentiated Protean stuff 'out there', but, as anti-realists often enough tell us, that is hardly a world worth saving.[4]

Within geography this relativist impulse has taken a number of forms in recent years. Under the rubric of subjectivism, for example, geographers have variously turned to idealism, existentialism, phenomenology, or linguistic philosophy in their search for a means of reinstating the human subject at the centre of the geographical stage. In all of these the subjective world of lived experience takes precedence to one degree or another. In some cases it takes the form of an appeal for the study of perceptions and human states of consciousness as of fundamental importance in human geography; in others the concern is to incorporate within geography emotional responses, not just cognitive ones, to environment and milieu; in still others there is the much grander claim that geographers need to transcend the normal limits of language and model their writing on figures like James Joyce because truth cannot be expressed in conventional linguistic forms.

There are, of course, many implications to be drawn from these programmatic statements of geographical anti-realism, but one of the most significant is the call for new ways of thinking about meaning, reference, and truth. One of the underlying reasons for this is the simple if awesome claim that the terms we use simply do not refer to existing entities in the world. Our language may help us to cope better with circumstances; but there is no cognitive fit between the world and our talk about it. The implications are enormous. We can give up the need to find direct empirical connection between terms and objects in the world; we can view knowledge not as presenting the world in some correct way, but as just helping us to get along in it, or to change it. Truth, to repeat, has nothing to do with accurately representing, or, as Rorty has it, mirroring reality; it is just, according to Rose, 'what we are well advised, given our present beliefs, to assert'. The purpose of geography, then, is not to tell us about how the world 'really' is: it tells us nothing about regions or landscapes or economic structures or human agency, for these are mere linguistic fictions; it is just the search for 'the right vocabulary, the right jargon, the best discourse in which to pursue the kinds of account which help us in the most basic sense, decide what to do'.[5] Colossal!

What are we to make of this radically pragmatic conception of truth and meaning? To be sure, we have no methodical or agreed way of settling disagreements about claims to knowledge or about warrants for making assertions; but why should that lead us to the radically anti-realist conclusion that there is no truth to be had? For the relativist, what it boils down to is this: truth is just what our peers will let us get away with saying. Truth is nothing more than what a scholar's circle will let her claim as warranting assertion, given the standard procedures of that community. This means that if I assertively claim that 'the duck is on the lake', the truth of my claim depends, not upon how things are in the world, but upon whether or not my peers will agree that I am using language acceptably. Surely there are problems with such a way of proceeding. For one thing, schemes of this sort run the risk of self-referential incoherence. If our peers will not let us get away with saying that 'truth is just what our peers will let us get away with saying', then the claim is not true. Why? Because it is what they will allow that passes as truth. This means that if the original claim is true, then it is not true![6] Besides, if the claim is that there are no privileged discourses that, so to speak, cut the world at its joints, then the anti-realist one is not privileged either, and has no cognitive claim to be telling us any truth. Out and out relativist schemes of this sort always run the risk of relativizing themselves. Of course this is not to advocate any sort of naïve realism about our knowledge of the world. There may well be difficulties in sustaining a realist account of, say, quarks or social classes; but it is hard to see why we should let these cases lead to the conclusion that all claims to knowledge are similarly problematic.

In broad outline, then, naturalism and humanism are two of the most pervasive ways by which geographers in recent decades have sought to conceptualize the relationship between the human and natural worlds. However, as we have already hinted, these ideas are to be found in various coalitions. Christianity, for example, could be seen as transcending these positions. On the one hand humanity is conceived as part of the stuff of nature, made from the dust of the ground, and therefore continuous with the rest of the natural order; but, on the other, the human species is to be regarded as bearer of the *imago dei*, and therein finds a dignity that raises it above the mere animal and asserts its independence and agency. Again certain versions of neo-Marxism seem perched between naturalism and humanism: for if indeed people make history, but not in conditions of their own making, then the dynamic of historical change is to be located in the interplay of structure and agency. For the Marxist, to be sure, humans are through and through part of nature; but they can take control of the world through acts of self-assertion and will-power. Whether this is a confusion or a clarification of the philosophical issues at stake cannot be adjudicated here.

It is our contention that the notion of the behavioural environment and its continued use within human geography can best be seen in the context of these fundamental philosophical questions. This is not to imply that the idea was originally conceived as contributing to the solution of problems of precisely this sort; it is even doubtful whether such issues could have been definitively articulated at that time. However, it is to suggest that the concern to integrate mind and nature, so central to the behavioural environment scheme, should remain high on the agenda of modern geographical enquiry.

The behavioural turn in human geography

The behavioural approach to geography was born essentially of the concern to identify those cognitive processes by which individuals and communities codify, react to, and recreate their environments. At least in part, the motivation was to get inside the heads of human actors in order to understand their behaviour in the world of external reality. As Harold Brookfield put it: 'Decision-makers operating in an environment base their decisions on the environment as they perceive it, not as it is. The action resulting from their decision, on the other hand, is played out in a real environment.'[7] Quite fundamental to this whole behavioural project, then, is the idea that a crucial distinction is to be drawn between the real world – the world as it is in and of itself, and the world as perceived – that is, the world as we humans take it to be. What are we to make of this claim?; and what are the implications to be drawn from urging an ontological bifurcation between the real and the perceived? Does an espousal of this dichotomy commit us to any particular philosophical position, or to any special methodology? Does it force on us either a positivistic or a humanistic perspective?; or is it sufficiently flexible to accommodate a variety of philosophical angles on the world? These are the sort of questions to be borne in mind as we reflect on the evolution of the behavioural environment theme in geography.

When William Kirk first introduced his behavioural environment schema in 1952, geography received its initiation into the relevance of Gestalt psychology for environmental knowing and human spatial behaviour. The concerns that underlay Kirk's turning to the Gestalt theorists were precisely those that we have just been considering, namely the need to find some creative approach to the relationship between humanity and nature. As he put it:

> What we need is some working hypothesis in which nature and humanity are brought under one discipline, and this can be found in the field of Gestalt Psychology. The theory of this school of thought developed from research of Max Wertheimer on stroboscopic motion and through the writing of Koffka, Köhler and others, has invaded the world of the humanities without leaving much impression on the geographical 'bridge' although many of its tenets find close parallels in geographical thought.[8]

The fundamental claim of the Gestalt movement, in opposition to atomism and behaviourism, was the notion of a 'Gestalt' as 'essentially an organised whole whose parts belong together as opposed to being juxtaposed or randomly distributed'.[9] In particular this holistic notion was believed to be relevant to the psychology of perception and much research was focused on the application of the theory to vision. Various laws of vision were forthcoming from the Gestalt theorists, the details of which have been thoroughly contested; but the notion of a perceived world as distinct from the 'real' world – an inexorably Kantian notion – found its way into geography. Indeed, this is scarcely surprising given the fact that Koffka himself drew a sharp distinction between what he called the 'behavioural environment' (the perceived world) and the 'geographical environment' (the external source of retinal stimulation).[10] Subsequently the Gestalt psychologists conducted various experimental studies of the problem-solving strategies of apes and humans, on the perception of colour, and on pattern recognition; and alongside these empirical investigations they made a spirited assault on the prevailing scientific orthodoxy of the day – positivism, as Koffka himself styled it. This was largely because positivism seemed to allow no place for the categories of meaning and value – the very factors that constitute the organizing principles of the behavioural environment. What Koffka and Köhler were interested in, in other words, was the means by which knowing subjects make sense of, or render intelligible, visual stimuli. When we interpret, say, a facial expression as communicating anger or sadness or joy we are attributing meaning and value to a particular physiognomic pattern. It is thus easy to see why Köhler should entitle his William James lectures of 1938 'The place of value in a world of facts'. For him, physical science too contributed to the understanding of the geographical environment; but the behavioural environment had its own storehouse of facts – facts constituted by values – which remained of special significance for the psychologist. For it is only the attribution of value that renders visual patterns intelligible; and it is therefore not surprising that this schema had implications for the constitution of the different sciences.

The idea that the mind contributes unifying and differentiating structures to the perception of the world was the fundamental notion that Kirk distilled from the work of the Gestalt theorists. By asserting the existence of an 'internal environment' – the behavioural environment – Kirk believed that 'the gap is closed between Mind and Nature'. Initially he reflected on the significance of the idea for historical geography, urging on his colleagues the task of reconstructing 'the environment not only as it was at various dates but as it was observed and thought to be, for it is in this behavioural environment that physical features acquire values and potentialities which attract or repel human action'.[11] It was plain to him, however, that the notion had far wider applicability. He himself always considered, for example, that the history of geographical thought was essentially the history of successive behavioural environments; and others have widened its range of application too. In this volume, for instance, Wreford Watson turns to the urban history of four major Scottish cities to show how their architecture, structure, and morphology were shaped by people's prejudices and perceptions. John and Margaret Gold look at the way in which behavioural environments can be manipulated, taking the case of the British architectural profession, some of whose members used particular photographic techniques to forge an adverse imagery of suburbia which in turn helped serve their own aesthetic and social interests. The theme of manipulation also comes through in Brian Goodey's scrutiny of small-town images where he looks at the relationship between the social and cultural meanings ascribed to town centres and the activities of urban designers.

The fact that the Gestalt psychologists rejected the philosophy embedded in logical positivism – essentially a theory about meaning – should not be taken to mean that Kirk found no time for the methods of natural science. Indeed, from time to time, as John Campbell's article in this volume confirms, he expressed his enthusiasm for a broadly positivist methodology, with all the naturalistic overtones that that suggests.[12] Kirk's conception of geography as an *environmental* science, his enthusiasm for the identification of 'natural' regions, and his unease about the voluntarist nature of possibilism, as Campbell makes clear, were all consonant with this outlook. Whether such a stance is intrinsic to his schema or merely contingent, of course, is a quite separate question; but that it was not of necessity incompatible with the notion of the behavioural environment is understandable. In principle a case could be made out for the legitimacy of interrogating the behavioural environment by the methods of natural science. To be sure Kirk never articulated a clear-cut procedure for reconstructing past behavioural environments. But if, say, *Verstehen* was proposed as a *modus operandi*, it would be entirely possible to use that strategy merely as a means of throwing up hypotheses that could be subjected to the normal testing procedures of natural science. Thus, any particular explanation for, say, an historical event could arguably come under some approved covering law. So in the same way it is entirely possible to argue for idealism as a means of 'getting inside the heads of past actors' while remaining committed to a positivist procedure of theory verification in history. To use an idealist *method* to investigate the past clearly does not imply that reality owes its structure to the activity of mind or that material objects do not exist. To adopt a *methodological* idealism, then, does not commit one to any particular *ontological* or *epistemological* stance.

In some ways, then, the possibility of marrying together an interest in environmental cognition with positivist social science helps to cast some light on the ways in which behavioural geography has developed over the past few decades. On the one hand there have been those who have turned to a mathematized psychology in order to scrutinize cognitive processes. Within human geography the search for the mechanisms underlying environmental perception, decision-making processes, mental maps, and so on was commonly cast in quantitative terms. Hence there developed the interest in information processing as central in explaining spatial patterns of behaviour. Indeed, this tradition is represented in this volume in the essay by Terence Anderson, a student of Kirk's, who resorted to the techniques of Repertory Grid analysis in his attempt to understand residential preference in Belfast. On the other hand there have been those who have increasingly expressed dissatisfaction with the positivist direction in behavioural geography, claiming that it ignored the subjective world of emotion, feeling, and value by focusing too exclusively on the mechanics of cognition – in short, bypassing all those distinctively human attributes that energize behaviour. These alleged shortcomings were all of a piece with an increased philosophical awareness, and a host of alternative stances ranging from idealism to phenomenology have been lassoed and put to work for geography. Again in the present collection aspects of this essentially humanist tradition find representation. Leonard Guelke, for example, turns to the classical tradition of Kantian idealism to emphasize that 'reality' is not given; it is constructed by mind. Geographers, therefore, must direct their attention to the 'internal' world of human consciousness in order to understand human behaviour in the world. Edward Relph, on the other hand, looks to phenomenology as a means of helping us to encounter the world as it is, not as mediated through some perceptual lens: for reality, to him, is not to be thought of as something that underlies or is behind perceptual appearances – it just *is* what we perceive. The immediately experienced world must therefore take the centre stage in geographical self-education. Yi-Fu Tuan, in his meditation, rejects reductionist accounts of human communication, emphasizing the ways in which language transcends mechanistic stimulus–response models, and reminds us that human creativity finds expression in silence no less than in symbol.

Irrespective of the precise historical influence of Kirk's original papers, what is remarkable is that representatives of both these traditions – let us call them 'naturalism' and 'humanism' – have found Kirk's project attractive. This is hardly surprising, however, given Kirk's self-confessed aim of bridging the gap between mind and nature. More, it is this very concern that renders Kirk's notion of the behavioural environment still worthy of contemplation. Indeed, as Douglas Pocock's essay reminds us, the behavioural environment provides one way in which the polarity between environmental reductionism and cultural autonomy may be transcended, and points to the need for a more integrative conception of the relationship between humankind and environment. To be sure, much philosophical spadework would have to be done if such a project were to be rendered ontologically and methodologically coherent. Still, the problems inherent in the naturalistic and humanist (or anti-realist) accounts of humanity's place in nature – with which we began – suggest the continuing need to find some means of transcending them, or at least of mediating between them.

Facing real issues and keeping good instincts

That the notion of the behavioural environment is still deployable in unravelling some of the philosophical tangles that confront modern geography is not, of course, to deny the difficulties and ambiguities in the original concept itself. Indeed, some of our contributors point to the problems, both empirical and theoretical, in the task of reconstructing behavioural environments. Drawing on his own experience of field-work in South-East Asia and Oceania, Harold Brookfield points to the dangers of researchers imposing their own values on the behavioural worlds of other peoples, the problems of conceptualizing the relationship between individual and group perception, and the practical consequences of inadequate or inaccurate or, worse, manipulative reconstructions of behavioural environments. Others like Chris Philo and Anne Buttimer sense the need to reach beyond the original formula and exploit the insights of post-foundationalist philosophers and social theorists. For Philo, the elucidation of the discursive contents of behavioural environments must engage with the hermeneutic project of Michel Foucault, who, while emphasizing the primacy of the 'document' in encountering the past, insisted all the while that former discourses are never just a simple window on the past. In Philo's view, Foucault's work provides one route between the extremes of objectivism and deconstructionism. Buttimer's concern, too, is to contextualize knowledge, in her case the geographical knowledge held by geographers. For her the social and cognitive uses of metaphor provide a means of giving structure to the diverse behavioural environments of practising geographers. Still others, like Ron Johnston, want to humanize Kirk's essentially naturalistic behavioural environment to understand the twin processes of community fragmentation and self-identification. The 'us–them' polarity is also the central point of focus in Michael Romann's investigation of the ethnically diverse perceptions among young people in Jerusalem, where cognitive dissonance may ultimately have the direst of consequences. To understand such processes, Johnston turns to the work of psychiatrists who have explained how individual and group self-assertion goes in tandem with the avoidance or negation of others. Making sense of these impulses towards segmentation and distancing will require interrogating behavioural environments to ascertain whether they are 'natural', or strategic or manipulative.

It must be conceded, then, that the behavioural environment is a rather elastic notion that has been used by partisans to justify a range of philosophical stances. Whether this is a strength or a weakness can only be determined in the long term. Suffice to note that positivists and humanists have found in it just what they each want.

An arguably more serious question lies in the area of ontology.

Does the behavioural environment actually refer to an existing entity – namely, is there an 'internal environment' which we can speak of as existing? Here the issue is whether realist status can be attributed to the concept or whether it is simply an instrumental notion that helps understanding. Allied to this problem, in addition, is the question of whether in using the terms 'behavioural environment' and 'internal environment' we are speaking literally or metaphorically. To conclude that we are essentially using metaphorical talk, of course, would not immediately address the ontological question. That would depend on resolving a number of other issues: is it necessary to translate a metaphor into literal talk before we can speak of any entity as existing?; in the case of the behavioural environment, is such a translation possible?; are there truths that can only be expressed in metaphorical language? In each case we would need to determine whether the idea is expressed in metaphorical form and, if it is, whether the metaphor is merely decorative or essential to the schema. Either way, the notion could be problematized in much the same way as has the idea of the 'mental map.'[13]

Even if we grant that there are such things as behavioural environments – and whether they exist at the level of individuals, communities, or both, remains an open question – there are fundamental methodological issues to be faced. Are there methodological tools available for getting inside the heads of actors? What they are and how the results are to be tested is certainly not spelled out in Kirk's schema. Indeed, given the critique of the deconstructionists in literary theory the idea of simply 'reading off' intentionality from observed actions or recorded texts is highly contested. This, at least in part, is because the decoding of signs, whether they be linguistic or social or environmental, is inexorably the activity of interpreters – an activity by which the interpreter makes meaning and fabricates facts. In a real sense, then, to the deconstructionists, texts do not exist until they are encountered and what matters in the encounter is what the individual makes of them. Whether we conceive of the texts as written records, cultural artefacts, urban landscapes, or whatever, the same hermeneutic task is engaged. The deconstructionists, of course, have not had it all their own way; but they have certainly made the 'reconstruction' of intentionality far from simple. Whatever the implications of this hermeneutic tradition for human geography actually are, its claims must clearly be taken on board by behavioural geographers.

Irrespective of the difficulties that still need to be faced in operationalizing the behavioural environment idea, Kirk displayed several good instincts in the thesis that he advanced. Right from the outset, for example, he sensed the need to widen geography's domain to incorporate human consciousness and cognition, namely to take seriously the way in which individuals and groups structure and make sense of their world. And yet this realization never pushed him towards an altogether autonomous subjectivism. There was truly a 'world of physical facts' and a 'world of social facts' that did not evaporate into the mists of fantasy or fabrication. Thus, he always viewed with grave suspicion, for example, the geographical interrogation of literary fiction as an end in itself. To be a concept-user, therefore, was not to be a world-maker – that metaphor was all wrong. Rather, it was to be a world-viewer.

Herein too may lie the reason why Kirk never allowed the incipient relativism of the behavioural environment notion to spill over into that relativistic extreme that renders the very notion of truth senseless. Thus, his belief that the same empirical data may arrange themselves into different patterns and have different meanings to people of different cultures never brought him to a radical scepticism about the possibility of actually reconstructing past geographies. It surely meant that naïve empiricism must for ever be banished from geography's kingdom; but it held out hope that a truer picture of humankind and its place in the world could be achieved.

To Kirk the integration of mind and nature perhaps understandably expressed itself in his uncompromising commitment to the unity of physical and human geography, and hence in his resolute insistence on geography as an environmental science. However, we do not need to share that particular rendition of the geographical tradition to remain committed to the necessity of keeping the objective and subjective worlds – the worlds of fact and value – bound together. In our own day this as often as not takes the form of a call to keep human agency and social structure in conceptual tandem. Accordingly the notion of the behavioural environment would alert us to the dangers in interpretative systems that give explanatory privilege either to structure or to agency, to naturalism or to humanism, to environment or to culture, to nature or to mind.

To be sure, the behavioural environment idea as originally formulated cannot be conceived as providing philosophical solutions to particular geographical problems. As we have said, it does not imply any specific method for interpreting, interrogating, or reconstructing behavioural environments, and still less does it offer criteria for warrants to knowledge claims about perception. What it does is to remind us that we are subjective agents in an objective world, and to warn us that to underestimate the human in geography is as dangerous as to overplay it.

Notes

1 R. J. Chorley, 'Geography as human ecology', in R. J. Chorley (ed.), *Directions in Geography* (Methuen, London, 1973), p. 158.
2 This analysis owes much to discussions with the American philosopher Alvin Plantinga.
3 See, for example, S. Rose, 'The roots and social functions of biological reductionism', in A. Peacocke (ed.), *Reductionism in Academic Disciplines* (Society for Research into Higher Education and NFER-Nelson, Guildford, 1985), pp. 24–42.
4 So, for example, N. Goodman, *Ways of Worldmaking* (Hackett Publishing Co., Indianapolis, 1978).
5 C. Rose, 'The problem of reference and geographic structuration', *Environment and Planning D: Society and Space*, vol. 5 (1987), p. 104.
6 See the analysis by A. Plantinga, 'How to be an anti-realist', *Proceedings and Addresses of the American Philosophical Association*, vol. 56 (1983), pp. 47–70.
7 H. C. Brookfield, 'On the environment as perceived', *Progress in Geography*, vol. 1 (1969), p. 53.
8 W. Kirk, 'Historical geography and the concept of the behavioural environment', *Indian Geographical Journal*, Silver Jubilee Volume (1952), p. 158, and reprinted as Chapter 2 in this volume. It was thus entirely understandable that Kirk should consistently maintain the unity of physical and human geography. Thus, in a taped interview with Rachel Regan, he described himself as a 'Unitarian' geographer. See R. A. Regan, 'British geographers on British geography', unpublished B.Sc. dissertation, Department of Geography, University of Birmingham, 1987.
9 T. R. Miles, 'Gestalt theory', in P. Edwards (ed.), *The Encyclopaedia of Philosophy*, vol. 3 (Macmillan and Free Press, New York; Collier-Macmillan, London, 1967), p. 318.
10 It should be pointed out that Kirk used the term 'geographical environment' to incorporate both his 'phenomenal' and 'behavioural' environments.
11 Kirk, 'Historical geography and the concept of the behavioural environment', p. 159.
12 This is also borne out by some marginal jottings that Kirk inscribed on the manuscript of a paper by D. N. Livingstone and R. T. Harrison, 'Immanuel Kant, subjectivism, and human geography: a preliminary investigation', which was subsequently published in *Transactions, Institute of British Geographers*, New Series, vol. 6 (1981), pp. 359–74. When the authors made reference to the humanistic rejection of the positivist approach, Kirk tersely scribbled 'Not me'.
13 See R. M. Downs, 'Maps and metaphors', *Professional Geographer*, vol. 33 (1981), pp. 287–93; E. Graham, 'Maps, metaphors, and muddles', *Professional Geographer*, vol. 34 (1982), pp. 251–60.

Chapter two

Historical geography and the concept of the behavioural environment*

William Kirk

'The world seen through the eyes, that is the prison house.'

The modern study of geography was born of three main disciplines – that of the great naturalist travellers studying at a time when Lamarckian 'needs' were being added to Darwinian 'Natural selection'; that of the political economists whose interests lay in the physical basis of political units and power; and that of those historians who were dissatisfied with a History which wandered on a formless earth. From such a birth the geographers of to-day have come into an inheritance at once rich but of an exceedingly complex nature and as specialisation has increased the problem of maintaining some centralising theme or purpose has become increasingly difficult. There appear to be as many geographies as geographers. The frontiers of study are being extended constantly into new realms and as with the expansion of the Roman Empire voices are raised from time to time asking for a halt to be made and a territorial demarcation drawn. However, almost invariably the debate on definition assumes the problem to be one of material content and resolves itself into a struggle for priority among certain groups of phenomena natural and human. A solution along these lines appears to be inherently impossible for on the one hand to maintain that all facts are geographical facts is tantamount to saying that no facts are geographical facts, while on the other to delimit a group of facts and refer them peculiarly to a geographical science seems equally absurd. A coconut palm is a fact in many realms of study. A true distinction can be made only when one thinks in terms of significance or values. The coconut palm acquires some additional character as it stands in the field of

* Reprinted, with minor bibliographical changes, from the *Indian Geographical Journal*, Silver Jubilee Souvenir and N. Subrahmanyam Memorial Volume (1952), pp. 152–60.

observation of the economist, the botanist, the geographer, or the peasant farmer in whose gardenland it is growing. This acquired character depends not so much on the act of observation as on the different modes of thought, levels of experience and sets of values implicit in the various perceptions. Thus the following discussion attempts to focus attention on function and methodology rather than on content, on the processes of geographical thought rather than on its material bases.

In his presidential address to the Geography Section of the British Association for the Advancement of Science in 1948 Lord Rennell of Rodd considered that the function of geography was to act as a 'bridge' between the humanities and natural sciences, as otherwise 'there can be no understanding but only an accumulation of facts.' This indeed is an ambitious challenge but nevertheless has far-reaching implications, not the least of which is that geographers must be prepared to deal with ideas and problems in a two-way traffic, in a manner intelligible to both ends of the bridge, and as with the transport of commodities to create value in transit. These ideas are many and have had a long history. First vaguely formulated in the cultures of the Fertile Crescent, they were pointed for the western world in the speculations of Hellenic thought, and eventually developed in the systematic studies of ancient Alexandria, that great clearing-house between west and east. For our immediate purpose the geographical consequences of but two of these ideas will be considered.

(1) **The Idea of Natural Law.** That is that 'measure of regularity or of persistence or of recurrence' of which Whitehead speaks,[1] without which there remains 'a mere welter of detail with no foothold for comparison with any such welter in the past, in the future, or circumambient in the present' and which is essential in the drive towards a methodology. In the growth of thought from the savage observance of the capriciousness of an unordered environment to the recognition of patterns in time and space various doctrines can be recognised. There is, for example, the doctrine of Natural Law as Immanent according to which the identities of pattern in the mutual relations of things arise from the elements they have in common in their fundamental nature; of Law as Imposed which presupposes an external source of order, flowing in an unchangeable course from the prime conception of a Creator who is sometimes presented as the Supreme Mathematician; and Law as Description when the essential criterion becomes the 'observed persistence of pattern in the observed succession

of natural things'. It will be seen that geographical thought contains elements of all three doctrines. We are interested in the nature of things, we have faith that the apparent capriciousness of certain phenomena is due to lack of knowledge rather than essential lawlessness, but at the same time we contribute to the Positivist doctrine and proceed by the scientific method of announcement of observed correlations of observed fact. 'I describe the earth' is still a prime responsibility of our discipline but the description is one of relationships expressed in patterns. Mukerjee defines ecology as the 'pattern of patterns' and in this feature geography and ecology are on common ground. The medium used by geographers to record such patterns is the map but as this is restricted in the main to the demonstration of spatial relationships, diagrams are introduced to show patterns in time. As our study grows from being a science of distribution to a science of human action in a multiple environment the latter increase in importance.

(2) **The Idea of Causal Relationship.** This is a development of the first idea in that in the search for relationship one set of facts or sequence of events is shown not only to correlate with another set but to be the cause of them. In their attitude to this idea the sciences and humanities have trod variant paths. To Helmholtz in the 19th century, as to Newton, 'The final aim of all natural science is to resolve itself into mechanics'. The Law of Causation was triumphant. Given a complete knowledge of State 'A', State 'B' could be deduced, from 'B' State 'C' and so on to the end of the world. During the 20th century, however, the development of the Theory of Relativity and the Quantum and research into the structure and behaviour of matter in its smallest units have reintroduced the idea of a certain looseness in nature. The Law of Causation is shown to be a statistical concept, the result of working with averages in a macroscopic world, while in the microscopic world chance and fate seem to take a hand, the steady progression of cause and effect appears as an illusion, and the Law of Indeterminancy takes its place beside that of causation. In the study of Man and the relationship of mind to material things changes of outlook have also occurred but often in the reverse direction. Thus Berkeley and the Idealists could declare 'All the choir of heaven and the furniture of earth have not any substance without the mind. So long as they are not actually perceived by me, or do not exist in my mind, or that of any other created spirit they must either have no existence at all or else subsist in

the mind of some eternal spirit'; but modern schools of thought have arisen in which mind can in no sense be distinguished from matter, and in which the former is described in terms of the latter. The Behaviourists, Pavlovian experimental psychology, the search for the physical springs of human action and personality, the conception of man as the sex cell's way of making another sex cell, the study of climate in its direct effects on the man-organism, the influence of diet and study of the process whereby soil becomes a poem, the study of disease and the secretion of the endocrine glands in an internal organic environment, the power of the 'genes' and 'mores', the dependence of individual behaviour on the pattern of culture of the group to which he belongs,[2] all seem to suggest that 'We are determined down to the last hair on our eyebrows ... could we know all the forces operating at any one moment of history we could predict the rest of history from that cross-section of time.'[3] The relevance of these counter ideas to geographical thought becomes evident if one believes that geographers should essay more than description. Indeed under the titles 'Determinism' and 'Possibilism' they have long divided students of the subject into rival camps and have been as much a stimulus to the heart of our study as correlation has been the key to our methodology. It is perhaps true to say that, at our present level of knowledge and techniques, the concept of 'Possibilism' claims most adherents. The story of man in relation to environment is presented as a conquest of nature, and with a 19th century enthusiasm in the technical achievements of mankind we believe that the road from palaeolithic savagery has been a glorious liberation from environmental controls, a growing potency of freedom of will. Of recent years, however, doubts have been expressed in the accuracy of this presentation and make necessary re-examination of the problem in a new form.

To handle such ideas and problems no static form of Geography will suffice. The search for Law through Relationship has led out the mind from the known to the unknown, not only in space but also in the fourth dimension of time, and the study of Historical Geography has grown as an integral part of the geographical discipline in response to this need for a space–time continuum in which the forces of human and natural action can be observed and measured. To restrict the horizon of the study in time would be as meaningless as the drawing of a frontier of content, for the present exists only as the moving integrate of the past, while to limit the geographer to a description of the stage and the historian to the human drama played out upon it would destroy the essential unity of the whole and sacrifice reality on the altar of academic convention. Stage and actors are in dynamic relationship both in space and time, and since the geographer is denied the method of the controlled experiment it would be folly to close the doors of the laboratory of the past. Historical Geography is no specialised compound of the facts of history and the facts of geography, whatever they may be, but an essential element in all geographical thought. Its objectives are those of geography itself – 'To grasp and reveal at each instant of their duration the complex relations of man, the actors in, and the creators of history with organic and inorganic nature and with the many factors in their physical and biological environment.'[4] Its methods of approach towards this objective, however, vary from school to school and a classification of these can now be made and their validity examined.

First – **the assessment of physical environment as a factor in History** – includes the many attempts to demonstrate the dependence of historical events upon physical occurrences of varying complexity, and usually proceeds by the system of correlation. At one end of the scale are those physical 'accidents' such as the storms which wrecked the Spanish Armada, the rains that 'rained away the Corn Laws', the movement of fishing grounds, the bumper harvest which quietened a revolution. At the other end the more complex rhythms of physical change, such as the progression of climate and vegetation following the Ice Age in the Northern Hemisphere, the cyclical variation in the carrying capacity of the Central Asian grasslands, the alterations in the ozone of the air, the sun-spot cycles and many others beloved of Ellsworth Huntington.[5] In addition there are those subtle, long term influences such as the physical personality of a region, geographical position, insularity, continentality, and peninsularity. In most instances, however, it is causal relationship which is sought and it is this feature which has brought the approach severe criticism from the sociologist, economist and historian. Toynbee, for example, in his great study of the geneses and growth of civilisations finds that 'neither race nor environment as hitherto envisaged has offered, or apparently can offer, any clue as to why this great transition in human history occurred not only in particular places but at particular dates.'[6] Different human responses spring from similar physical challenges, and to any environmental problem at least three alternative answers are possible for human groups, viz., to move away from the problem, to stay and surrender, to stay and conquer by changing either themselves or the problem. As soon as one descends from the broad generalisations of average conditions to particular cases and situations the same disturbance of cause and effect as met by the scientist in his transition from a macroscopic to a microscopic world faces the historical geographer. The same alternatives are open to him as to a human group faced by an environmental problem – to move his approach, to resign himself to possibilist interpretations, or to restate the problem in more precise terms.

Secondly – **the analysis of the evolution of landscape** – Landscapes like other phenomena can be described in terms of 'structure' and 'process'. 'Structure' involves the idea that it consists of identifiable parts which are organised in functional relations and work together as wholes, while by 'process' is understood the action of forces either external or internal which result in change. 'Structure' is the static, 'process' the dynamic aspect, but the former however stable it may seem exists only in terms of process, while the latter is inconceivable without and works through structure. Thus it is that many historical geographers have concentrated on describing the process by which the structure of present cultural and physical landscapes have evolved. Two stages are discernible in this approach. First the recognition of the elements present in a particular landscape, their abstraction and analysis by what the historian knows as 'lines of development'. To this we owe the many fine works which have appeared on the growth of patterns of settlement, the development of the city-idea in its many forms, the pioneering spread of the village unit, the changing significance of the cross roads and market, the fascinating topic of field patterns, the growth of trade and the means of carrying it on, the changing value and exploitation of mineral resources, as well as those referred to as geomorphological. Geomorphology indeed is a vital and indispensable part of the discipline of historical geography. Would that the precision usually evident in geomorphological research was also applied to the other structural elements in landscape. Having taken to pieces the second stage is the observance of the reassembled wholes in their evolutionary aspect – the analysis of regionalism in a fourth dimension. Space does not allow one to dwell on the thorny question of the concept of the region, except to remark that in the opinion of the writer the immense labour which has been expended in the delineation of regions on the basis of uniformity of enclosed phenomena, while useful as an aid to memory and the writing of textbooks, has shown little profit in the advancement of our study. It is an endless task which suffers from the same handicaps as schemes for the classification of men – that with more and more subdivision to find the perfect unit the conclusion eventually reached is that every point is individual. Of greater value is the search for regions of process, regions in time, to demonstrate the changing pattern of forces operative in the creation of landscape, at times approaching a balanced equilibrium, at others in a state of flux when the introduction of some new element disturbs old harmonies. The bond of participation in process is stronger than similarity to the eye.

Thirdly – **the consideration of man as an agent of environmental change.** This is obviously a special case of the second approach and the opposite pole of the first. During the past half million years the work of men has not been without its effect on the physical environment. The great works of reclamation, the moving of mountains, the clearing of forests, the watering of desert places, the creation of soil are but a few instances recorded by historical geographers of the influence of men in natural processes and the disturbance of the concept of natural climax. But the increasing momentum of human change, and the increased capacity of man as a tool-using animal to alter his habitat is double edged. Through the pages of William Vogt[7] man stalks like the central figure in some gigantic Greek tragedy, with a power for destruction keeping pace with an increasing power for creation, threatened by self-annihilation and the loss of much good in the eradication of a little evil.

Fourthly – **the reconstruction of past geographies.** In this approach a cross section in time is taken and subjected to geographical analysis. The earth is seen as some huge photographic plate exposed periodically to record the patterns made by living things, then, as the archaeologist lays bare by excavation the details of the past occupation of a site, so the historical geographer by the same delicate removal of superimposed patterns can strip backwards in time and reveal another 'present' for analysis without drastic alteration in his methodology. The method has many attractions but carries with it the problem of recognition of varying rates of change, and the establishment of 'synchronisms' between natural and human phenomena. How often in historical works does one see the attempt to explain an ancient cultural landscape in terms of a 20th century physical landscape. The history of Europe or the Ganges Basin for example cannot be understood without the reconstruction of its primitive vegetation. Fortunately, in recent years, under the general title of geochronology, a number of techniques have been developed in which the earth itself is being called upon to reveal a time scale of physical change against which human change can be synchronised.[8] The analysis of pollen from trees long since dead, the rhythm of climatic change as recorded in the nature of tree rings, the work of de Geer and his followers on the sequence of varve deposition, the greater understanding of the stages in the withdrawal of the ice sheets in Europe and America and their parallels in sub-tropical latitudes, research into the formation of river terrace and fluctuating coastlines, a deeper knowledge of plant succession, make possible the replacement of a relative by an absolute chronology, and together with written evidence, when available, facilitate the drawing of a more accurate picture of past physical environment. The work of J. G. D. Clark on the food-gathering peoples of Northern Europe during the early post-glacial period bears witness to the potentialities of the method for the historical geographer as well as the prehistorian.[9] Another danger of this method of approach to which the geographer is more prone, however, is the reference of 20th century motives to the action of human groups in the past, a danger not unlike that undergone by Western Geographers when they attempt to analyse the geography of the East. The task of getting into the minds of the folk who built the Buddhist cities of S. E. Asia is of the same order as that of revealing the motives of those tribal groups who built the 'henges' and barrows and left their mark on the downlands of England. What was really in the mind of the Saxon farmer on the pioneer fringes of Dark Age Britain, and in what ways did his assessment of values and motives differ from those Aryan villagers who colonised the 'dark and trackless forest' of the Middle Ganges? For upon such questions as much as the environment they met depends the final outcome and the reality of the cross section geography.

We have now raised more problems than we can hope to answer in this short paper – the need to seek values as well as facts, to function as a bridge of ideas, to develop a four dimensional study, to concern ourselves with action and process as well as distribution, and to restate the problem of environmentalism in some new form. From our consideration of the methods of approach in Historical Geography at least one thing becomes clear, however, and that is that if we are to make any headway towards the 'enunciation of law' we must find some answer to the creeping paralysis of possibilism. The concept of the Behavioural Environment is tentatively suggested as a step in that direction.

Possibilism springs from the fallacy of regarding environment and mankind as two entities of unreconcilable character, the processes of the former conforming blindly to eternal laws, the processes of the latter dominated by the freedom of will. In other words, we are still thinking in terms of 19th century thought. What we need is some working hypothesis in which nature and humanity are brought under one discipline, and this can be found in the field of Gestalt Psychology. The theory of this school of thought developed from research of Max Wertheimer on stroboscopic motion and through the writing of Koffka, Köhler and others has invaded the world of the humanities without leaving much impression on the geographical 'bridge' although many of its tenets find close parallels in geographical thought.[10] A 'Gestalt' is almost precisely what we imply when we define a region in its dynamic aspect, a whole which is something more than the sum of the parts; the psychology of Mackinder, his ability to see things in functioning wholes, to alternate figure and ground when examining a map, and his sense of 'requiredness' are essentially the attributes of the Gestalt Psychologist;[11] but here we are concerned especially with what is termed the psycho-physical field. A simple illustration of this can be demonstrated from Figure 2.1.

Figure 2.1

This diagram can be viewed either as a black vertical cross on a white background, or a white diagonal cross on a black background, and with concentration the two can be made to alternate with an impression of movement. Now this movement does not take place in the physical environment of this page but in the mind dependent on the stimuli of light passing through the eyes. In addition the very fact of seeing a pattern of crosses rather than a series of unrelated lines implies that forces are at work in the mind which tend towards an organisation of environment into 'figure' and 'ground'. Moreover the act of concentration to cause the pattern to change implies the existence of another source of energy – the 'self' of the reader – who has acted in this way because his cultural heritage includes not only the ability to read but to read an English script. If at some future date he sees a similar diagram he will probably repeat the process owing to the momentum of the pattern in his 'memory'. Thus environment, light energy, the chemical state of the brain, past and future mental states and the action of the individual observer are bound by a pattern of forces which originate in what must be termed a 'psycho-physical' field. Action in fact results from the attempt to relieve stresses in this field – it is easier to see and remember a pattern than otherwise. Translating this into geographical terms with the aid of a second diagram [produces] Figure 2.2.

If 'A' represents the physical environment including both the

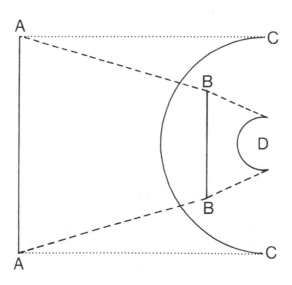

Figure 2.2

physical and cultural landscapes, and 'C' represents the physical human group or individual present in this environment, the physical state of 'C' will depend in part on the character of 'A', but any action of 'C' in this environment will commence in the relief of stresses in an internal environment 'B' which are as much the product of the group culture 'D' as the act of observation of the physical environment. This internal environment we may call the 'Behavioural Environment' and in this environment the gap is closed between Mind and Nature.

In as much as in Historical Geography we are concerned with the behaviour of human groups in relation to environment it behoves us to reconstruct the environment not only as it was at various dates but as it was observed and thought to be, for it is in this behavioural environment that physical features acquire values and potentialities which attract or repel human action. Certain features may entirely change in significance in different Behavioural Environments. A piece of ground of little value in the Behavioural Environment of one group may become suddenly desirable in that of another group. For example what was the Behavioural Environment of Stamford Raffles when he chose the site of Singapore and took such pleasure in the sight of the Union Jack moving above the mounds of 'the old city'; or of the Factors at Surat when Bombay was acquired from Portugal by a government at home who in the words of Lord Clarendon thought of it as 'the island of Bombay with towns and castles therein, which are within a very little distance from Brazil' while Antonio de Castro,[12] Governor of Goa, could opine that 'India will be lost the same day in which the English nation is settled in Bombay'? Such inquiry would certainly help the historical geographer to avoid the trap into which the Persian Governor Megabazus fell when a century after the foundation of the two Greek city colonies of Chalcedon and Byzantium, upon hearing that the former was planted on the Asiatic side of the Bosphorus seventeen years before the Byzantines built their city on the European side remarked 'Then the Chalcedonians must have been blind men all that time.'[13] For the aspect of the Behavioural Environment which had attractive value at the time of the foundation of the cities was agricultural not commercial and in this respect the former site was superior. Furthermore the concept is of value in that it places emphasis on process and action in real situations, the problems of real life, such as that which faced the Roman generals in their military occupation of Britain when no Ordnance Survey map was at hand to plan the strategy of history textbooks, when an otherwise well informed contemporary could write 'Britain, the largest of the islands known to us Romans is so shaped and situated as to face Germany on the east and Spain on the west',[14] and the best of contemporary map makers could turn Scotland through 90°, and display only a rudimentary knowledge of 'capes and bays' geography.[15] Finally the analysis of Behavioural Environments should throw new light on the question of environmentalism and overcome the weaknesses of the 'environment as hitherto envisaged'. For it may be observed that the greater man's knowledge of environment, the greater his awareness of its potentialities, the greater not less does it influence his actions. This influence is not direct as envisaged in the possibilist theories but operates through and is transformed in the Behavioural Environment. The story of man may yet be written not as a conquest but as a marriage.

Notes

1 A. N. Whitehead, *Adventures of Ideas* (Penguin, Harmondsworth, 1948), p. 130.
2 See among others: R. Benedict, *Patterns of Culture* (Routledge, London, 1935); W. B. Cannon, *The Wisdom of the Body* (Norton, New York, 1939); H. S. Jennings, *The Biological Basis of Human Nature* (Norton, New York, 1930).
3 V. H. Mottram, *The Physical Basis of Personality* (Penguin, Harmondsworth, 1949), p. 121.
4 P. Vidal de la Blache. [The editors have been unable to locate the precise source of this reference.]
5 See in particular E. Huntington's use of correlation in *Mainsprings of Civilisation* (John Wiley, New York, 1945) and *Civilisation and Climate* (Yale University Press, New Haven, Conn., 1915).
6 A. J. Toynbee, *A Study of History*, abridgement of vols. I–VI by D. C. Somervell (Oxford University Press, Oxford, 1947), p. 59.
7 W. Vogt, *Road to Survival* (Gollancz, London, 1949).
8 For a discussion of these techniques see A. Zeuner, *Dating the Past* (Methuen, London, 1946), and for a partial application of the method in Asia see H. de Terra, 'The quaternary terrace system of Southern Asia and the age of man', *Geographical Review*, vol. 29 (1939), pp. 101–18.
9 J. G. D. Clark, *The Mesolithic Settlement of Northern Europe* (Cambridge University Press, Cambridge, 1936).
10 For a general statement of 'Gestalt' theory see W. Köhler, *Gestalt Psychology* (Bell & Sons, London, 1930) and K. Koffka, *Principles of Gestalt Psychology* (Kegan Paul, London, 1935).
11 In particular see the treatment of the Atlantic Ocean as a unit in Sir H. J. Mackinder, *Britain and the British Seas* (Heinemann, London, 1902).
12 In a letter to the King of Portugal dated 5 January 1665.
13 *Herodotus* (Heinemann, London, 1920–5), book IV, ch. 144.

14 Tacitus, *Agricola* (Heinemann, London, 1932), ch. 10.
15 For an analysis of Ptolemy's map of the British Isles see: I. A. Richmond, 'Ptolemaic Scotland', *Proceedings of the Society of Antiquaries of Scotland*, Fifth Series, vol. 8 (1921), pp. 288–301; H. Bradly, 'Ptolemy's geography of the British Isles', *Archaeologia*, vol. 48 (1883–4).

Part two

Reflection

Chapter three

The concept of 'the behavioural environment', and its origins, reconsidered

John A. Campbell

> There have been many definitions of this modern geography. I do not propose to try to add another, nor to indulge in verbal niceties about its scope. For there is an alternative method of appreciating its content, and that is not to enumerate what it contains, but to examine the intellectual environment in which it has grown up. (H. C. Darby)[1]

> Géographie humaine. ... Prenez garde. Tant qu'il s'agissait d'une géographie 'tout court' comme dit Vidal, s'occupant naturellement et largement de l'homme ... – point de difficultés, point de conflits, point de périls. Du jour où on a prétendu créer de toutes pièces une science autonome baptisée géographie humaine; du jour où on a ainsi introduit officiellement l'homme dans la place, de ce jour les difficultés sont nées. Philosophiques, si je peux dire, et méthodiques. (L. Febvre)[2]

Most references in the geographical literature to Kirk's first definition of the behavioural environment, and his later elaboration of it,[3] form part of retrospective (and often introductory) surveys of the 'behavioural revolution' and its origins,[4] and, as such, readily and legitimately attract the allegation of Whig history written backwards to justify a later orthodoxy.[5] Whether the tumult was in fact of epic proportions, its consequences for good or ill, or indeed who can be seen as its chief instigators, will not be of primary concern here, for what such talk or rumour of revolution all too easily obscures are the historical circumstances which moved Kirk to write his two major papers, and account for the arguments they contain. What will be attempted instead, therefore, is, first, an assessment of these two early papers and of others by him, exemplifying their conceptual thrust, against the background of the problems and challenges facing geography in the immediate post-war years; and, second, some consideration of subsequent developments in geography in the light of what Kirk had hoped might happen, and as a context for his own later contributions to the continuing, and increasingly lively, debate among geographers. By thus working fore and aft in time, it is hoped that the concept of the behavioural environment will be seen in a perspective truer than that gained from that history-with-hindsight productive of nothing so much as a crick in the neck.[6]

Because of its connotations of sharp break with the past, enormous upheaval, and ensuing novelty, among the most unfortunate consequences of currency of the term 'behavioural revolution' is that it obscures at least some 'revolutionaries'' preoccupation with issues of long standing, not to say perennial interest, among geographers. This is especially true of Kirk's contribution; and of greatest significance in this respect has been his frequently stated, and evidently paramount, objective of finding a satisfactory epistemology for the entire discipline of geography. While others have further concealed this objective by their emphasis on 'behavioural geography as method'[7] (sometimes verging on psychologism), and/or by their regrettable segregation of a field of 'behavioural geography' (or 'perception studies'),[8] it is plain that Kirk has always regarded his epistemological endeavour as of supreme importance. This is clear, for example, from the very title of the 1963 paper, its serial review and rejection of the existing definitions of geography, and its conclusion:

> one must not forget that an older, and perhaps more funda-
> mental, way of defining the fields of particular disciplines is to
> ask, not what materials they study, not what techniques they
> use, but for what kind of problems in human experience have
> they been invented to provide answers.[9]

Intimately related to this search for a more secure epistemological niche for geography is Kirk's anxiety to achieve an effective response to another enduring problem of the discipline, the preservation of its unity, not only against the traditional dualisms, but also in the face of a rapidly increasing specialization that presages terminal fissiparism. As some attributes of the schema he suggests as a solution to these epistemological dilemmas appear to have been overlooked in later discussions, they warrant some preliminary comment.

Notwithstanding his repudiation of geography as the science of all 'earthbound facts' or a 'particular category of earthly facts',[10] it is all too evident that it has been Kirk's unswerving aim 'to restate the problem of environmentalism in some new form'.[11] The student of Stamp and Wooldridge, the veteran of the wartime Burma campaign,[12] the field worker in physical as well as human geography (and their marchlands),[13] he assumes it to be almost self-evident, virtually beyond doubt that geography is an environmentalist discipline. For him, accordingly, it remains both the great glory and weakness of modern geography that its 'multiple origins' are traceable to the 'heady period of intellectual ferment in Europe and Victorian England in the latter half of the nineteenth century'.[14]

One intended corrective to the drawbacks of this intellectual inheritance is the concept of 'phenomenal environment', a concept implicit in his 1952 paper, though not formalized until 1963, and one overly neglected due perhaps to its later introduction, and lesser innovativeness, than the concept of 'behavioural environment'. Meant essentially to counteract the often static conception of physical environment among many traditional human geographers, historians, and sociologists, the concept of phenomenal environment emphasizes the stage's dynamism (as a result of the cumulative impact of physical processes, and human agency, over time), and obtrudes prominently in Kirk's frequent reminders of the importance of environmental reconstruction, and of the increasing battery of techniques that make such retrieval possible with growing accuracy. As a concept, however, phenomenal environment is not without difficulties, even flaws, and this may also help to account for the seeming reluctance to adopt and employ it. Inasmuch, for instance, as 'organic processes and products (including human populations)' are assigned to the phenomenal environment, and as it claimed that 'areal variation in the physical constitution of mankind is no less a fact of nature than the ecological complexity of equatorial rain forest', then there would appear to be more than a hint here of humanity dehumanized.[15] Indeed, still more revealing in this regard is an earlier reference to 'the physical environment including both physical and cultural landscapes'.[16] And if one were to speculate as to a likely reason for Kirk's commission of these naturalistic errors then a plausible source would seem to be Wooldridge, his commendation of Huxley's *Physiography*, and the case he outlines for a 'natural geography', in which 'man comes inevitably into the picture and in his right place as part of the terrestrial unity'.[17]

As for the concept of 'behavioural environment' only some prefatory remarks will be offered at this point. Unquestionably, much of the persuasiveness of Kirk's presentation of this concept derives from its graphic demonstration, whether the mandala of the first paper, or the sketch by Toulouse Lautrec in the second, and the alternate ways in which they can be envisaged. In part, however, these graphics are a misleading guide to his thought, for whereas they are unchanging, and their alternate construal depends entirely on the observer,[18] Kirk rightly insists, as we have seen, on the reality of an actively changing, not static, phenomenal environment, and far from advocating cultural determinism, he stresses the two-way interaction between phenomenal and behavioural environments. Indeed, he urges an acceptance that 'a learning model lies at the heart of geographical consciousness',[19] and quite consistently, therefore, in a manner simultaneously reminiscent of Huntington and Toynbee, he can at times appeal to the phenomenal environment to act as a 'stimulus' to provoke an observed social 'response'.[20] Indeed, the ability of a society's culture to preserve a legacy of such primordial encounters is a theme present in much of what he has written, notably in relation to erstwhile forest frontiers.[21]

It is clear, furthermore, that between 1952 and 1963 Kirk slightly altered his original diagram of environments, and their relatedness, to provide for greater direct interaction between individuals/societies and their phenomenal environments. While the lines designated A–C in 1952 remain essentially in their original positions, and continue to provide for the direct impact on those environed by the phenomenal environment, the positions of the lines in 1952 designated A–B are altered (compare Figures 3.1a and 3.1b). The resultant reduction in the ambit of the 'behavioural environment' allows greater provision for those kinds of direct environmental impact which comprise some of the subject matter of 'physical anthropogeography'.[22]

Finally, it could be argued that by lumping together 'development of geographical ideas and values' with 'awareness of environment', the former a kind of encoded folk memory, Kirk preserves the coherence of the concept of behavioural environment, but at the considerable sacrifice of thereby rendering it unwieldy. That this is unnecessary is apparent from Bird's exposition of the relevance to geographical science of the philosophy of Karl Popper. Geography, like any other science, Bird maintains, is concerned in essence with the content of, and the interactions between, the 'three worlds' distinguished by Popper:

> World 1 encompasses all the phenomena in objective reality 'out there'; World 2 is the subjective world of individual

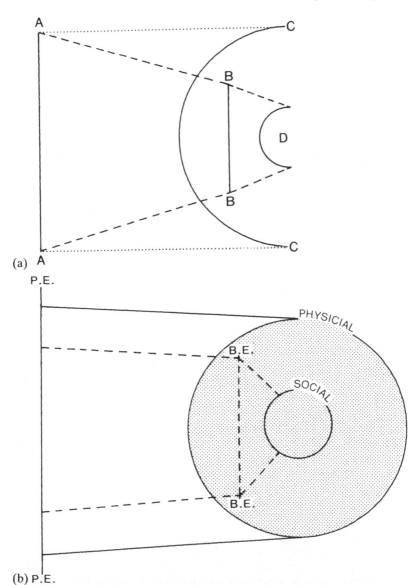

(a)

(b) P.E.

Figure 3.1   Different models of the behavioural environment
(a) Source: William Kirk, 'Historical geography and the concept of the
   behavioural environment', *Indian Geographical Journal*, Silver
   Jubilee Volume (1952), p. 159, fig. 2
(b) Source: William Kirk, 'Problems of geography', *Geography*, vol. 48
   (1963), p. 366, fig. 5

consciousness, inside our heads; and World 3 is the world of objective knowledge enshrined in concepts coded in some form of language.[23]

Bird then goes on to show that, while Kirk's 'phenomenal environment' is the equivalent of Popper's World 1, his 'behavioural environment' is coextensive with both Worlds 2 and 3. This latter equivalence therefore suggests the possibility, on grounds of both logic and elegance, of conceptual fission, but Kirk might resist a separation of this kind, as not only would it further fragment his overall schema, but it might also impair recognition of forms of interaction between Worlds 1 and 3.

In sum, therefore, the concept of behavioural environment now seems less inclusive, actually and potentially, than it was at its inception, and in turn this means it is less able to lend that coherence to geography which it originally appeared to promise. Whereas in 1952 Kirk could aver 'in this environment the gap is closed between Mind and Nature',[24] the seal has come since to appear less hermetic; and surely symptomatic of this is his introduction in 1963 of the overarching concept of 'geographical environment', which an uncharitable commentator could regard as emergency provision to forestall an incipient rupturing of conceptual sutures. While obviously designated to keep the entire conceptual edifice upstanding, it is again a concept that has found little favour; and this is hardly surprising, for on the one hand its embrace is so vast as to vitiate its serviceability,[25] while on the other realization of the need to restrict its scope by the qualification 'geographical' inserts a tautology that further compromises its worth.[26] It is apparent, however, from more recent papers, that Kirk himself retains confidence in his schema, and its key environmental concepts, and if still further proof were needed of the persistence of his epistemological concerns, then the subtitle of his presidential address to the Institute of British Geographers in 1978,[27] 'towards a geographical philosophy', is proof enough.

As Kirk himself has stated, his preoccupation *ab initio* with the philosophy and unity of his chosen discipline cannot but be seen in the light of British geography's fortunes in the immediate post-war period. Lecture theatres full of ex-servicemen with experience of various, far-flung theatres of war were often taught by young, newly recruited lecturers, like Kirk, with similar first-hand, hard-won acquaintance with distant foreign parts; and as, a little later, 'the first hints of expansion in the university world were becoming apparent', the still larger departmental staffs then possible permitted subdisciplinary elaboration, but fostered rather than endangered the integrity of the discipline itself. At Aberdeen in the 1950s, Kirk recalls:

> We had our specialisms in research but our objectives were to produce graduates with a broad command of the subject. We had our debates on the nature of the discipline and in a precocious fashion ushered in both the quantitative and behavioural approaches to the subject long before they became fashionable elsewhere – but as approaches to a unified subject, not as conflicting dominant philosophies. Weber, Von Thünen, Lösch, and company were common grist to the teaching mill long before the appearance of the texts of the Cambridge school on the analysis of spatial patterns.[28]

What emerges clearly, therefore, is a condition of intellectual liveliness, and not that state of quietude which assertions of a subsequent abrupt paradigm shift, 'the quantitative revolution', would lead one to presume. If further evidence were needed of the vigour of the current debate, then the founding in 1954 of the albeit short-lived journal, *Geographical Studies*, the self-proclaimed mouthpiece of the rising 'third generation' of modern British geographers,[29] is all the evidence necessary. The appearance in its pages of several papers reopening the environmental determinism/possibilism controversy was part of a more widespread examination of this hoary old chestnut in geographical circles; and it would be an error not to appreciate that it was also as a contribution to reinvigorated discussion of this philosophically important issue of free will versus necessitarianism that Kirk proposed his concept of the behavioural environment. For it was by more precisely specifying the components of what he came to call the geographical environment, and their interrelationship, that he hoped human geography would be able to rid itself of what he described acidly as 'the creeping paralysis of possibilism'[30] – an allusion later quoted, with apparent approval, by Hartshorne.[31] His immediate aim seems to have been akin to that of Spate at the time – namely, 'to break away from the besetting danger of a sterile and lazy empiricism to reach a reasonable modus vivendi between the two schools'.[32] Kirk's ultimate objective, however, was the formulation of a worthwhile body of theory, with the associated feasibility of prediction, without which no science worthy of its name he felt was possible; and undoubtedly, as his attack on possibilism hints, it was by guaranteeing that the 'Ge' was not taken out of Geography, but instead found appropriate conceptual accommodation, that he, like Spate, believed this could best be done.[33] Indeed, it was through Spate's good offices that Harold and Margaret Sprout were introduced to Kirk's work, and took account of it in their essay 'Man–milieu relationship hypotheses in the context of international politics' and their subsequent book *The Ecological Perspective on Human Affairs with Special Reference to International Politics.*[34]

While, more recently, Spate has repudiated as 'outmoded ... the old view of geography as primarily a study of man–environment relations',[35] and has left the discipline to practise as an historian, the trajectory of Kirk's thought – and, of course, career – has remained resolutely environmentalist. Appreciation of this has not been helped by mistaken characterizations of him as a 'non-positivist'[36] which discount both Kirk's own professions to the contrary,[37] and ways in which he has applied his ideas in practice that make it all too apparent. Indeed, instances can be found where there seems to be little, if any, trace of behaviouralist thinking, and instead recourse to earlier naturalistic concepts like that of 'natural region', widely expounded, for example, by his teacher, Stamp. Thus, in his account of the Damodar valley, Kirk seems fully prepared to employ Gestalt as a holistic metaphor where the phrasing of the argument is essentially in natural-system terms:

> The catchment basin is the vital natural region, if one defines a natural region not so much on the basis of similarity, which is a generalized myth in any case, but of *process*. It can be considered a unit in which, according to the second law of thermodynamics, the interplay of forces strives toward an equilibrium in which stress and energy are at the minimum and into which come human groups to disturb or sustain the natural processes. It can also be considered a *Gestalt*, in which the whole is greater than the sum of the parts, and yet in which the alteration of one part affects the whole.[38]

In this, as in other cases, such statements also probably reflect the kind of epistemological concerns to which reference has been made already; and pertinent to this point too is the fact that elsewhere Kirk's survey is sometimes at a macro scale, and working with a 'broad brush' could be regarded as encouraging a more emphatic environmentalist treatment.

It is instructive to compare Kirk's reformulated environmentalism, and the circumstances of its origins, with the thinking of other human geographers who, in the aftermath of the Second World War, likewise expressed their dissatisfaction with the legacy of

Vidal's 'environmental possibilism'.[39] In the German tradition, for example, the thought of Schwind shows quite remarkable parallels with that of Kirk: there is that same appeal to psychology as a solution to geography's philosophical dilemma, in the case of Schwind to the 'Umweltlehre' of Jakob von Uexküll;[40] that same emphasis on cultural landscapes, their structures, and transformations, as in essence 'mentifact' derivatives, as ultimately the outward manifestations of human mental activity; and that particular interest in Oriental cosmologies, as illustrated, for example, in Schwind's exposition of the landscape impress of Confucianism, Buddhism, and Bushido, in the *Geisteskultur* section of his account of people and landscape in Tokugawa Japan.[41] Indeed, it is when Kirk considers the cosmological foundations of traditional urban planning in South Asia that he departs most obviously, and most explicitly, from the more naturalistic environmentalism to which he holds in much else he has written; and if this is accepted as a not unfair judgement, then one might add that nowhere in Kirk's work can be found that kind of concise, manifesto-like statement which Schwind has produced, pointing up the insidious pervasiveness of culture, and the symbolic dimensions of space.

In France, meanwhile, the post-war years were similarly 'le temps de craquements',[42] and here once more the limitations inherent in the ecological tradition, constraining the proper future development of human geography, were the subject of intense discussion, and a springboard for subsequent work in social, urban, and political geography. Conscious of the restrictiveness, especially in modern urban-industrial societies, of naturalistic concepts, like *genre de vie*, central to the Vidalian position,[43] and committed to an ideology in which the dualism of people and nature was fundamental, George was only one of the more prominent French Marxist geographers urging a changed philosophical stance, and utterly dismissive of the merest suggestion of a recrudescence of environmental determinist thought.[44] Despite many similarities in the reasons for their discontent, it is more worthwhile to note the considerable differences between George (and his fellow Marxists) and Kirk in the solutions they worked out. Whereas the former did not feel that his position fatally compromised the unity of geography, the latter was later sharply critical of those Soviet geographers whose ideology led them too to espouse dualist views. Yet in indicating the source of this dualism Kirk points to that very issue which in much of his own geography he seems often to have resolved, 'behavioural environment' notwithstanding, to the detriment of a truly humane geography:

They argue that not only are the materials of human geography ('economic geography' in the U.S.S.R.) completely different from those of physical geography, but since social processes are unlike natural processes, the laws governing the one cannot be applied to the other.[45]

Though free from the radical ideology of his Marxist contemporaries, the writings of another post-war leader of French geography, Gottmann, share their impatience with tradition, and also contain guidelines for the refashioning of human geography. Fresh from his wartime sojourn in the United States, he published in 1947, for example, 'De la méthode d'analyse en géographie humaine', in which he asserted that in its research methods physical geography was more assured than its human counterpart; that interpretations of human behaviour in terms of physical milieu could only be crude, more intuitive than deductive, the product of least scholarly effort; that humankind is superior to the rest of animal creation precisely because it is within human power to reshape nature; and that as a concept 'le genre de vie est surtout un outil de description, description raisonnée, bien sûr, mais où l'explication ne fait encore qu'accompagner et soutenir la description sans pouvoir s'en dégager et moins encore la précéder.'[46] Sensitive to the dynamic nature of human geography's subject matter, he goes on to counsel the need for a more process-orientated approach, suggesting that human geographers in the Anglo-Saxon tradition are more alive to this need than his fellow countrymen. Where thereafter he indicates the significance of American ethnographers' studies of cultural patterns, and even more so when he considers 'l'aspect le plus purement humain: le facteur psychologique',[47] the convergence with Kirk's then current thinking is readily apparent. Beyond this point, in many (though not all) respects, the parallels cease, however, because, to a far greater extent than Kirk, Gottmann voices greater reservations as to the potential benefits the physical and, still more so, the natural sciences can confer henceforth on studies of process in human geography. And the corollary would also seem to apply, that Gottmann's humanistic commitment is both more consistent and far stronger than Kirk's as the following passage suggests:

En géographie humaine, la méthode d'analyse doit toujours tenir compte de ce facteur spirituel, rechercher le ferment psychologique, en apprécier la force. Il faut là nous écarter un peu de la logique des expérimentateurs, surtout les naturalistes; Claude Bernard a pu dire: 'Le fait juge l'idée', car nous souhaitons toujours des preuves factuelles. Mais c'est l'idée qui suscite le fait; indirectement sans doute et en ouvrant la porte à bien des surprises, mais la vie psychologique est à la base du dynamisme de la géographie humaine (et nous englobons en ce sens, dans 'humain', l'économique comme le politique et le social). Une méthode d'analyse, pour être scientifique dans notre discipline, doit donc renoncer à un matérialisme géographique trop simple pour les faits et admettre que les forces spirituelles peuvent compenser des pressions dont l'énergie puise à d'autres sources.[48]

While the New World spirit of 'Promethean'[49] endeavour Gottmann envisages as fundamentally responsible for the creation of Megalopolis can be regarded as a secular counterpart to the cosmological underpinnings of Oriental urban design explored by Kirk, the latter's political geography lacks any formal equivalent to the concept of 'iconography' that forms so critical a component in Gottmann's conceptual analysis of the political geography of states. Could the reason be that Kirk's attention has been directed primarily not to political cores but to militarized peripheries; and there to instability rather than stability? Or does the reason lie deeper, in Kirk's more naturalistic, less humanistic sympathies, for, as Gottmann avers, though normally by nature conservative, iconography can periodically become a factor for change: 'It is a powerful factor and it fights for stability and resists change, except for short periods in every nation's life when claims for change – i.e. for expansion – are raised, supported by a special interpretation of the iconography.'[50]

Mention of cores and peripheries leads naturally to a discussion of Kirk's thinking on the idea of the region. Just how hotly disputed had the debate about regional geography become by the early years of his professional career can be gathered from the trenchant criticisms contained in 'The inadequacy of the regional concept',[51] an essay by Kimble, a fellow student of King's College, London. Kirk clearly shared many of Kimble's reservations about past practice in regional geography, and judged a need henceforth to 'search for regions of process, regions in time, to demonstrate the changing pattern of forces operative in the creation of landscape.'[52] In making this search, it is equally clear that Kirk believed that his own thought could be of help, for he wrote: 'A "Gestalt" is almost precisely what we imply when we define a region in its dynamic aspect, a whole which is something more than the sum of the parts . . . [a] psycho-physical field.'[53] Although his subsequently frequent application to regions of the holistic Gestalt metaphor might seem to imply a humanistic treatment,

Reflection

however, closer examination of the matter and manner of what in fact he writes is often more reminiscent of the social physics of the latter part of the last century. Because the human geographer is denied the method of the controlled experiment, then it is to 'the laboratory of the past',[54] and its regional variety, that Kirk advises him to resort. As 'the fourth dimension of time'[55] is thus opened up, the *va-et-vient* of human affairs, played out in 'theatres of action',[56] is revealed as the caprice of 'forces', of 'structures' with an in-built tendency to internal 'differentiation',[57] now tending more or less to 'dynamic balance or equilibrium',[58] now sustained by 'geographical momentum',[59] sometimes 'cyclical' in their 'structural transformations',[60] sometimes 'catastrophic'[61] and irreversible, but withal impelled relentlessly, irresistibly, inevitably onward by an unwilled dynamic.[62] The result is the very opposite of what one might expect of a behavioural geographer: human initiative and striving seem to count for little, history appears changeless, and its geographical frame again essentially constant. Consider, for example, his observations on frontier making:

> An internal mechanism characteristic of most frontiers appears to be operative from the earliest times and if it can be demonstrated that similar resultant structures arise in various environmental circumstances then it would appear that we are dealing with a model that might provide a start for the development of general theory.[63]

Or again, his conclusions on 'Prehistoric Scotland: the regional dimension', where the region as constant emerges still more boldly:

> This ranging backward in time from the firmer regionalism of the proto-historic period into the more dimly perceived regionalism of the prehistoric millennia suggests a remarkable continuity in the regional framework of human behaviour in Scotland. Technological change, the development of new tools in stone, bronze or iron, the infusion of new ideas, or the coming of new people, may alter behavioural patterns from time to time and transform social structures, language, architecture and settlement forms, the quality of life, and the range and quantity of material possessions; but at certain scales of human adaptation to the environment, regional constraints and potentialities appear to exert themselves repetitively.[64]

It could be contended, of course, that elsewhere Kirk's version of the integrative functioning of processes, physical and human, over time, in a regional setting appears less naturalistic because by allowing for internal and external perceptions it conveys a greater sense of historicity, and of place. This appears so, for example, when, echoing Mackinder, he argues that sea-based perspectives are as important as the more customary land-based ones, and that geography urgently needs an adequate scheme of 'sea regions'.[65] Yet even in such instances, a static, single (often Eurocentric) regional classification is again invariably adopted, as in his early survey of '*the* Indian Ocean *community*'[66] (my emphases). When such accounts, and the ones just mentioned above, are set alongside those of Spate, now historian of the Pacific since Magellan,[67] the comparison yields insight, inasmuch as the latter reveal how different is a truly humanistic treatment. Candid in his admission that '"the Pacific" is a European artefact', Spate divides its histories into two sorts: first, '"The history of the Pacific", of the Ocean as an Ocean . . . so far dominantly a European history in the writing, and largely in the making'; and, second, '"The History of Pacific Peoples" . . . in practice "of Pacific Islanders" . . . not . . . quite so exclusively Euro-dominated, either in the events or the historiography.'[68] In future historical research, Spate wants not 'a dichotomous share out, the Islander taking the inner or local function, the European relating it to outside world forces and trends',[69] but participation by both in conducting and marrying the two. And not for him either, in his own monumental study, the lure of universalist, timeless history.

> Our Pacific is not the Mar del Sur of the 16th century, nor even the South Seas of the 19th; and in assessing the human response to its immensities, one must always consider the setting at the time: an essay in the difficult art of controlled forgetting.[70]

Admittedly, Kirk's interest in societies remote in both time and (sometimes) space creates enormous, almost insuperable problems in delineating the 'inner experience' of regional societies.[71] Sporadic archaeological finds make the reconstruction of regional personalities at best tentatively informed inferences. However, even where access would have been far easier for him, and the possibilities of representing the insider view infinitely more attainable, as in those coastal settlements reflective of 'the duality of social and economic life in the wider province of north-east Scotland', he describes only the binary settlement pattern of seatown below/farmtown above, and rest content we must with an all too brief tailpiece: 'The steep brae between divides two social worlds.'[72] At bottom, then, the root cause of Kirk's failure to realize the humanist potential of his own schema can only be attributed to his self-proclaimed positivism; and all the evidence would suggest, therefore, that he was and is not entirely in sympathy with Spate when, like others since, the latter claims that:

> in the last resort we can only grasp the essence of regions, or a region, by a geographical empathy akin to the historical empathy demanded in Croce's famous passage about the Ligurian peasant, and indeed some element of 'identification' seems to me to mark the best regional writing.[73]

As Kirk insists that his 1963 paper differs significantly from his earlier statement of 1952 in that the former's chief concern was with decision-making, then not only would it be reasonable to anticipate his subsequent clean break with naturalistic assumptions (which has not happened), but also to detect signs of this in his previous work, particularly in economic geography, given current changes in that subdiscipline of importance for the so-called 'behavioural revolution'. If, however, one takes his review of the cotton and jute industries of India, for example, and, despite the appearance here again of the Gestalt metaphor, it is to spatial economics in the neo-classical mould, with its naturalistic over-tones, that he in fact appeals:

> Rather than attempt to solve the locational problem of an individual firm Lösch investigates those regional structures which condition and motivate industrial location, and in so doing differentiates various types of agglomeration. His regions are defined not by self-sufficiency, factor mobility, or homogeneity, but by the complicated regional structure itself – the *Gestalt* – simultaneously and mutually determining its integral parts, while his theory of agglomeration is sufficiently general to embrace town growth and morphology as well as industrial sites. Lösch's theory thus supplements that of Weber, providing the general equilibrium to Weber's partial equilibrium . . . [74]

The absence here of conceptual scope for human agency is interesting in itself, but also against the backdrop, at the time this passage was written, of the widespread currency in British geography of the term 'functional region', for as employed it too often had the same deficiency. Although the term had contemporary equivalents in other traditions, a main British source was G. W. S. Robinson; and, most unfortunately, it was his key paper of 1952 that had the critical defect of rendering Vogel's term 'Zwecklandschaft', with its connotation of human purposiveness, as the ambiguous 'functional region'.[75] Indeed, some years later Pfeifer expressed his unhappiness with the term 'functional region', for he felt it could be severally interpreted and used, was 'zu mechanistisch', and made no clear provision for 'gewordenen kulturellen, die historisch bedingten Besonderheiten'.[76] And in expressing this opinion Pfeifer would not have been a solitary voice, for contemporary articles by fellow German geographers were likewise unequivocal on the cardinal issue of the distinctiveness of human intentionality, and its basis. A little earlier, for example, in an article notable for its historiographic depth, Lautensach showed his appreciation of the fact that 'die treibende Kraft in der Sphäre der psychischen Kausalität ist die menschliche Absicht einer zweckbestimmten Einwirkung auf die Natur oder auf die von früheren Generationen geschaffene Kulturlandschaft'.[77] Still earlier, Bobek had shown the same appreciation in one of the papers that established him as a leader of post-war German social geography:

> Sie wird vom autonomen menschlichen Geist bestimmt und kennt dementsprechend keine Kausalitäten, sondern nur Motivationen. In ihr gelten jene Regeln, Grundsätze und Gesetzmässigkeiten, die Soziologie für die Betätigung und das Wirksamwerden von Urteilskraft und Willen innerhalb von Einzelpersönlichkeiten, Gruppen und Massen herausgearbeitet hat. Die Geographie muss, wenn sie sich zur Wissenschaftlichkeit erheben will, von diesen Grundprinzipien Kenntnis nehmen und sie in das Kräftespiel der Landschaft einfügen.[78]

Kindred sentiments were not lacking in France, as we have seen, and as additionally can be gauged from Labasse's thesis on the organization of financial space around Lyon; and from his subsequent identification of a 'géographie volontaire/voluntaristic geography', received with such enthusiasm by Ackerman, a very early American proponent of a behavioural approach in human geography.[79] However, Kirk, it would seem, did not share that same strength of conviction displayed by some of his German and French contemporaries; for while he talks of the importance of 'conscious purposeful human behaviour' in space, and of the significance for *human geography* (my emphasis) of decision-making, he can also, almost within the same breath, appear to envisage geography *as a whole* (again, my emphasis) as a behavioural science, and contemplate both human spatial behaviour, and 'the behaviour of a rain drop', as simply falling within one and the same 'spectrum or range' of behaviours.[80] Such use of the word behaviour has very justifiably attracted recent criticism from certain behavioural geographers;[81] and, apart from his positivism, at least one further motive that can be suggested for this liberality of use in Kirk's case, as with his overworking of the term 'environment', is his ever-present wish to maintain the unity of the discipline, and to focus attention on recognizably 'geographical problems'.

Yet to be too quickly dismissive of Kirk's treatment of purposive human spatial behaviour would certainly be overhasty, if for no other reason than the role Kirk has assigned consistently to great men at the 'turning-points' – the sudden 'Gestalt shifts', the 'times of decision' – in history.[82] Among those so denoted by him have been such figures as Raffles, the Factors at Surat, Megabazus, and, of course, his commander in Burma, General Slim. What he says of each, and at length about Slim, is highly persuasive, but the case he makes does nevertheless also provoke scepticism, and chiefly on the following grounds. First, all the instances cited are either colonial or military, where great and often undisputed power has rested very largely in the hands of an élite coterie or of a commander-in-chief, and where a high degree of common purpose means that the exercise of power has occurred in a setting difficult to regard as entirely representative. Second, as Hartshorne rightly concedes,[83] geographers in the past, while not slow to acknowledge the geographical creativity of individuals, have shown far more reluctance to confront the philosophical issues inevitably raised by such acknowledgement. Writing this in Belfast, it is difficult to refrain from quoting Jones, for his frank admission on this score in his account of shipbuilder, E. J. Harland:

> Here, the force of personal initiative can hardly be overemphasized: the geographer's usual interpretations, involving site and resources, become woefully inadequate. As his job is to seek explanations wherever they may be found, the geographer must recognize the purely human factors which dominate the history of Belfast's shipbuilding at this point. He must look to the man rather than to the place, to foresight rather than to site, and to initiative more than to the accessibility of raw materials.[84]

While Kirk firmly believes that the role of great people should in future occupy a far higher position on the geographical agenda, whether his positivism will allow him to make those considerable philosophical concessions that its raising, and debate, might eventually suggest, must surely be in some doubt. For Samuels is quite sure as to why geographers thus far have side-stepped the philosophical implications of 'landscape biography' by great people: 'The reason for such neglect on the part of geographers is that Geography is itself a product of Enlightenment rationalism.'[85]

In historical geography, therefore, it is unsurprising to find Kirk treating geographical change, and the resultant structures, as the product of vast impersonal forces in a fashion that recalls the positivist ethnologists, sociologists, and historians of the late nineteenth century. Symptomatic are relics of unilinear evolutionary 'developmental stages', for instance, of 'the intellectual steps leading from primitive animism to all the world's great religions',[86] and it is a similarly tell-tale feature of his 1963 model of 'the cultural-spatial diversification of mankind via a series of decision taking situations' that historical change over time is calibrated according to a *single* standard of 'achievement',[87] thereby making no allowance for relativity in cultural values. Such an approach is the more readily understandable in one who moves easily from geomorphology to prehistory and on to historical geography, from the study of 'stages' and 'ages' and 'periods'. At first, his references to 'development cores ... transmitting pulses of energy outward through peripheral systems',[88] despite their almost Helmholtzian ring, seem innocuous enough, mere energetic analogies, but not when he proceeds to reify those 'core-periphery structures', to envisage them as 'entrenched in the landscape',[89] and endowed with a will of their own.

> A periphery *unwilling* to be drawn into the mainstream of national life and *resenting* penetration in the economic, social and, above all, the political field is a very different problem from a periphery *demanding* a greater share of the perceived economic benefits from development which appear to be concentrated unduly in a core [my emphases].[90]

Moreover, though he does maintain elsewhere that the 'concept of centrality or central place varies according to the behavioural system within which it is operative',[91] he does not always seem entirely sensitive to this himself, as when, for example, he describes thus the achievement of political unity in the British Isles between 1800 and 1920:

> After a thousand years of warfare, colonisation, commerce and anglicisation one after another of the Celtic provinces of *the* Atlantic *rim* had been formally united with *the* political *core* of the well populated English lowlands and granted the right to elect representatives to the sovereign parliament in London [my emphases].[92]

Elsewhere, he does, it must be said, allow that cores can in time become peripheries, and vice versa, but again the means he sometimes proposes smack less of the behaviouralist, and more of the nineteenth-century scientific materialist: '[spatial structures] are not eternal and regional inequalities characteristic of one developmental phase can be reversed by a change in technology or the discovery of a new resource'.[93] At other times, he is more judicious, as when he writes that 'no geographer would deny that potentials vary from place to place, nor that with changing technology and perceptions such potentials can vary over time',[94] but such cryptic statements still leave open and uninterrogated the question of *primum mobile*.

From historical geography springs Kirk's interest in political geography,[95] and, more particularly, the geography of frontier making. Deprecating the usual geographical practice of treating frontiers in a purely descriptive, classificatory way, he has long been an advocate of a more dynamic, genetic approach, in which frontiers become mobile interfaces, akin to the Burma battlefront – 'the edges of the perceptual system, the linkages and discontinuities in information networks that structure the system and the sets of perceived objectives that structure and give it "direction"'.[96] Traceable here are not only the intellectual fruits of his wartime experiences, but also the influence of his association at Aberdeen with the behavioural political scientist Frankel;[97] and the result is a fundamentally transactional view of politics, closely resembling that advanced lately by House, in papers and his book *Frontier on the Rio Grande*.[98] Remarkably, neither man refers to the expression of similar views by Wright in a paper of 1944, in which initial remarks about the immaturity of political geography and its being 'perhaps the most "human" phase of geography', are followed by the outline of a transactional model, with the appended observation:

> For 'political groups' we might substitute 'any human groups' in the diagram. The triangle would then suggest a wider realm of study, of which political geography would be only a part (though perhaps the most important part): 'the geography of active human contacts'. The immense field would embrace the whole great triangle of Men, Other Men, and Earth – a triangle as old and fully as perplexing as the proverbial one that is commonly called 'eternal'.[99]

What might make difficult acceptance of the transactional nature of Kirk's stance in political geography is the fact that his twin concepts of behavioural/phenomenal environments in essence provide for a subject/object relationship (even though he does allow, as has been shown, for humanity, albeit dehumanized, to form part of the phenomenal environment), and not one, to use

Wright's terms, entailing interaction also between 'Men' and 'Other Men'. This is indeed, it must be conceded, the major defect of the diagrams Kirk uses to illustrate his thinking. In practice, however, as the Mandalay paper shows clearly, his political thought is plainly transactional, and that this has long been so is further witnessed by what he has written of those 'inner Asian frontiers', set within 'the enormous buffer zone' of the Old World, an integral part of 'one of the most important political geographical features of the earth's surface'.[100] A sometimes noteworthy aspect of these analyses has been Kirk's explicit recognition of the reality of one contending party misreading the intentions of other parties in the historical drama, with all the ensuing unintended consequences that flow from such misperception. Thus, of the aftermath of the 1954 Sino-Indian treaty agreement, he writes:

> Subsequent events, however, have demonstrated that India's confidence in this treaty as a final settlement of possible points of dispute along her extensive northern frontier with a powerful communist state was misplaced. She has been too willing to accept at their face value Chinese promises that economic aid to Tibet would in no way detract from the regional autonomy of that country or the theocratic government of its Dalai Lama, and there is no doubt that the speed and militancy of the Chinese occupation of Tibet both surprised and shocked the Indian Government.[101]

Thus, it is beyond doubt that, although Kirk's phenomenal/behavioural environment schema does not provide for the kind of subject/subject interaction just described, he himself can show an awareness of the often momentous consequences of such conscious, albeit at times falsely conscious, human interactions.

Kirk's recognition of false consciousness, and its historical significance, can probably be regarded as just part of the intellectual legacy of his close contacts with Karl Mannheim, the author of *Ideology and Utopia*, and a modern founder of the sociology of knowledge, who taught Kirk at the University of London's Institute of Education in 1946. Apart from the great faith placed by both men in the key role education could play as guarantor of future democratic freedoms,[102] the most outstanding similarity in their thought lies in the strong appeal to both of the holistic configurationism of Gestalt psychology and, indeed, it appears likely that it was Mannheim who first introduced Gestaltism to the younger man.[103] Further similarities are not hard to discover. Of the significance of ideology in history, for instance, both were in agreement. Despite Kirk's professed positivism, the powerful sway of ideas and ideologies, past, present, and future, surfaces intermittently, if unpredictably, as a theme in his geography, an internal contradiction not wholly dissimilar to that between Marxism and Weberianism as currents in Mannheim's thought and work. Thus, drawing to the close of his inaugural lecture at Leicester, Kirk is resolute in affirming:

> It is clear that although each individual, each generation, and each government develops its own peculiar view of the world, as a result of geographical location and historical tradition, there are a number of major perceptions which have persisted over long periods and exercised great influence upon strategic thinking and political behaviour. Sometimes it has been a map, constructed on a particular projection, that has served generations of statesmen as the basis of political and military planning. Sometimes it has been a compelling theory of spatial relationships and historical causation that has moulded the viewpoint and action of political leaders. In so far as Geography is a behavioural science it is one of the tasks of the geographer to assess to what extent such 'pictures' or models of the world have influenced human spatial behaviour in the past and how far such environmental conceptions or misconceptions continue to exert influence on present political and economic decisions.[104]

And once again, on the subject of utopian visions, the two men exhibit the same doubts, believing such projects impractical, and unadaptable to the flux of historical circumstances.[105] To quote Kirk once more:

> No state ... can afford to dismantle and rebuild the ship to a new design while still on the high seas. It can seek to repair faults and weaknesses in the structure ... , but cannot effect total Utopian transformation of the whole without participating in its own destruction. Indeed achievement of Utopia, by its static perfection, would signify the demise of development ...[106]

The demonstrable pragmatism of this statement is, moreover, evidence of yet another similarity in the viewpoints of the two men, for, at least during his period of residence in England, Mannheim's thought too assumed this same Anglo-Saxon bias. And what, therefore, Coser identifies as the 'patent weaknesses' of this stance in Mannheim applies equally to Kirk's position:

> Judgements as to what contributes to adjustment and what does not are not only largely normative, but they are likely, at best, to be *ex post facto* judgements. It may be possible in

many cases to decide after the event which ideas contributed to historical adjustments, but it seems impossible to make such evaluations about contemporary ideas.[107]

In Kirk's case these criticisms are all the more unfortunate in so far as even those historical interpretations he offers which appear most clearly idealist, notably those of city morphology, relate often to ideal-typical blueprints – in Vitruvius' *De architectura* and in Kautilya's *Arthasastra*[108] – for anticipated conditions rather than to the historical realities of their actual implementation/non-implementation and functioning/malfunctioning, which could well have brought into play quite different, and at times perhaps conflicting, motives. By failing to direct attention to these motives also, and in effect thereby masking them, the scale and character of motivational interplay in the past, and thus the forces shaping successive geographies, is so unwarrantedly limited as seriously to risk major historico-geographical misinterpretation.[109]

If, however, major parallels exist between the thought of Mannheim and Kirk, then there are also striking contrasts, which help further to clarify the precise nature of Kirk's standpoint. First, whereas Kirk's original emphasis was certainly on percepts and not concepts, and on the external empirical evidence of behaviour rather than on motivation and its sources in concrete historical-social situations, Mannheim's position, at least in his German phase, is quite different:

> These persons, bound together into groups, strive in accordance with the character and position of the groups to which they belong to change the surrounding world of nature and society or attempt to maintain it in a given condition. It is the direction of this will to change or to maintain, of this collective activity, which produces the guiding thread for the emergence of their problems, their concepts and their forms of thought.[110]

Second, while Kirk's positivism, with its naturalistic presuppositions, underpins his belief in a unified geography, and in the absence of fundamental difference between the sciences of human-kind and nature, Mannheim is equally persuaded of positivism's error, and the consequent requirement for epistemological revision:

> What we have hitherto hidden from ourselves and not integrated into our epistemology is that knowledge in the political and social sciences is, from a certain position, different from formal mechanistic knowledge; it is different from the point where it transcends the mere enumeration of facts and correlations and approximates the model of situationally determined knowledge ...[111]

Third, Kirk, true to his positivism, appears to cleave[112] to the notion of a value-free social science, while Mannheim, spurning positivism, is equally convinced of the invalidity of such a notion. Fourth, and finally, Kirk inclines to a conservative ideology[113] (although with Marxist undertones in his polemic!), a stance different from Mannheim's 'planning for freedom', deliberate in its avoidance of both ideologies and utopias.

At least some of what Kirk did not absorb from Mannheim he subsequently assimilated from other sources, and a predominant influence has been the psychologist and philosopher, Piaget. Accordingly, Kirk has paid increasing attention to cognition, and to the formation and transformation of cognitive structures, and has placed ever greater stress on the fundamental importance of the learning process. If, however, Kirk's work is to form the touchstone for a truly humanistic geography, then not only were such changes imperative, but further changes of the sort Kirk himself envisages are also crucial; and some might suggest more changes in the direction of idealism that he might not wish to contemplate. Whatever the respective arguments, it will suffice here to recall Piaget's contention that Gestalt psychology alone could never provide a sufficient intellectual basis for any humanistic endeavour (geographical or otherwise) by virtue of its naturalistic pedigree.[114] To assert, therefore, as Kirk has done in the past, that Gestalt psychology provides an adequate 'working hypothesis in which nature and humanity are brought under one discipline',[115] and, more especially, a means of understanding how mental structures 'acquire values in cultural contexts',[116] is to misconstrue the original aims and concepts of Gestaltism's founding fathers. Apt to be critical of introspectionists' 'meaning theory', they were instead intent on showing 'that organisation in a sensory field is something which originates as a characteristic achievement of the nervous system'.[117] That socio-cultural influences on perception were not of the essence in Gestaltism, though not incompatible with its tenets, is abundantly clear from the following statement by Köhler, one of this psychological school's founders:

> sensory units may have acquired names and may have become richly symbolic in the context of knowledge, while existing, nevertheless, as segregated units in the sensory field prior to such accretions. Such is the conception which gestalt psychology offers to defend. It even goes so far as to hold that it is precisely the original organization and segregation of organized wholes which make it possible for the sensory world to appear so utterly imbued with meaning to the adult, because, in its

gradual entrance into the sensory field, meaning follows the lines drawn by natural organization. It usually enters into segregated wholes.[118]

If latterly, therefore, some have found it possible to level at Kirk's model the (not altogether legitimate) criticism that it 'does not specify by what processes this alleged distortion (of the external environment ... by a ... psychological representation of it) takes place',[119] then it is because that model is derived (imperfectly) from the work of psychologists whose particular emphasis was not on the socio-cultural but on the neurophysiological determinants of perception, as the Gestaltists' key concept of isomorphism amply confirms.[120]

Indeed, it should be remembered that some Gestalt psychologists have been quite open about Gestaltism's neglect of the social-cultural: Koffka, for example, in the penultimate chapter of his *Principles of Gestalt Psychology*, entitled 'Society and personality', frankly admits 'the incompleteness of the preceding discussion', and the 'sketchy'[121] treatment of the main problems of social psychology he then offers; and Katz, in the final chapter of his book, is still more forthright:

> On occasion the older psychology, oriented towards sensation, was reproached for giving too little weight to higher mental functions which were less closely connected with the senses. In this respect Gestalt psychology itself, so strongly influenced by the problem of perception, is hardly better off ... what is a symbol as seen by Gestalt psychology? Isomorphism fails to supply an answer.[122]

Thus, Kirk's inclusion of the realms of society and culture within his hypothetical Gestalt model finds little justification in the original conceptual basis of that school of psychology, and consequently the claim he makes for that model as a framework within which all things geographical can be fully embraced is not well founded. However questionable Kirk's Gestalt hypothesis, his very deployment of it does nevertheless demand close scrutiny for another, related reason: for its use serves incidentally to disclose notable differences between Kirk and his teacher Wooldridge in the manner of their representation of the geographical significance of human subjectivity. While both evince a very lively appreciation of how critical are human perceptions of natural environmental potentials, Wooldridge's work is patently idealist at source, socially and culturally sensitive, and concretely historical in its setting,[123] whereas Kirk's, Gestaltist and so naturalistic in inspiration, has been more formal and universalist, and somewhat lacking in socio-cultural reference. Indeed, it could be argued that the cumulative impact of Piaget's 'considerable'[124] influence on Kirk has been such as to allow the latter to build upon his original Gestalt framework in such a way as to shift his own position steadily closer to that of Wooldridge. This is because Piaget, like Kirk, takes Gestalt psychology as a starting-point, and for both it assuredly retains enduring merits, notably as the source of a still serviceable and insightful holistic metaphor, and for its concept of a structured cognition organized in adaptation to, and in equilibrium with, its environment. More pertinent, however, are what Piaget identifies as Gestaltism's defects, and how his thought surmounts these. One is that Gestalt theory is essentially static and ahistorical,[125] whereas empirical findings require, and he provides, a theory of active and progressive mental development: 'the most convincing evidence that cognitive development takes place and is dependent on an equilibration that is distinct from normal physical equilibrium is precisely that the operational structures of the intelligence are irreducible to perceptual "Gestalts"'.[126] Another attack he makes is that Gestaltism is individualistic, and discounts the social matrix within which learning occurs; thus, it is noteworthy that Piaget distinguishes three principal influences on the formation of cognitive structures: 'maturation of the nervous system, experience acquired in interaction with the physical environment, and the influence of the social milieu'.[127]

In consequence, Kirk owes to Piaget a more socially informed perspective, with a bias towards social interaction, and a more truly dynamic approach to change over time, sudden 'Gestalt shifts' giving way (by and large) to the more gradual, albeit staged, 'structural transformations'.[128] Yet the validity of applying a model of individual intellectual growth to the study of historical change remains, of course, a moot point; and doubts as to its correctness must surely increase when one recalls the limited role Piaget assigns to culture – as merely accelerating or retarding the developmental advent of stages he considers as at least empirically well established, and in all likelihood universal. Whether Piaget is right or not has still to be established, but his position does suggest a major reason for Kirk's conspicuous and continuing silence about ways of satisfactorily assimilating the dimension of culture within his overall theoretical schema.

A formative influence on Kirk's more recent thought, reinforcing Piaget's socially imbued psychology, and confirming its bias towards idealism, has been the 'field theory' of Kurt Lewin. Given its lineal descent from the Gestalt school of psychology, Kirk's appeal to this theory is unsurprising, and at the very least it gives added support to a holistic concept of society perhaps again derived originally by Kirk from Mannheim:

> Society at a given stage of its evolution is no mere agglomeration of exactly observable individual data, of sparse events and relationships all of which, added together, in some way produce the picture of the whole, but a combination of interdependent phenomena, and even more: a structured whole or Gestalt (a term used here in a general, not merely psychological, sense). If we divide this whole into its parts, and focus attention on the individual functioning of each part, then we shall necessarily overlook a very important aspect of the functioning of the parts, namely, their relation to the whole to which they belong.[129]

What field theory has to offer, however, is more than endorsement of holism, because what it also does is to sustain a Goffmanesque concept of society as the product of social interaction. Indeed, as shown above, Kirk has in fact long inclined to such a view of society; hidden, not revealed, by his diagrammatic schema, it has been implicit in his frequent resort to dramaturgical metaphor – in the idiom of Strabo, to regions 'as theatres of action'.[130]

On Kirk's own admission, his venture into field theory is still at an early stage, and he readily concedes the future need for more work in this respect. What is remarkable none the less is that nowhere does he link his allusions to topology with any to Lewin's closely related science of hodology, and this omission again can be regarded as symptomatic of Kirk's own position. Because, as Piaget notes, it was Lewin's chief contribution, under the influence of the neo-Kantian Cassirer,[131] to apply

> ... Gestalt principles, derived from work on perception, to the problems of motivation and personality. ... In Lewin's field forces not only the orientation of the subject played a role but the forces of other people and even of objects which had a valence – positive if they were attractive and negative if they did not.[132]

Now, as the absence of references to hodology might suggest, it is the essentially motivational aspect of Lewin's work that fails to emerge clearly when Kirk refers to it; and there are two particular reasons here that can be advanced to account for this suppression of motivation. First, notwithstanding the influence of Mannheim, Piaget, and Lewin, but true to individualistic Gestaltism, Kirk's outlook *de facto* remains only weakly and imperfectly social in perspective.[133] As already noted, many of his examplary hero figures are all powerful, often military leaders, striving for a total victory at all costs in which adversaries, and their strategies, are frequently lost; or, alternatively, they are heroes by design alone, people whose grand plans, not their implementation, are what he describes. This is scarcely the stuff of everyday life in which social position is the arbitrated outcome of more modest transactions between less grandiose actors or groups. Thus, it is not only by choosing to concentrate on 'larger than life' individuals in far from normal situations, but by neglecting ordinary social institutions, and the way they constrain, or facilitate, individuals' actions, that actuating motives, and their structural contingencies, tend to slip from Kirk's purview. Second, when Kirk urges on geographers the importance of adopting a dynamic 'locating' as opposed to a static 'locational' viewpoint, he would appear to provide ample scope for the play of motivations in the 'act of creation'.[134] Paradoxically, however, this fresh viewpoint is effectively obliterated by his pragmatic conception of behaviour as adaptive, for in common with all functionalisms this conception fails 'to make conceptually central the negotiated character of norms, as open to divergent and conflicting interpretations in relation to divergent and conflicting interests in society'.[135] Doubtless the basis of Kirk's stress on adaptation is to be found in his enduring environmentalism, but its confirmation by Mannheim in pragmatic statements made during his final, more practically orientated English period has also to be recalled:

> Whether the same reality (natural or social) presents itself in terms of a mechanistic total or of a Gestalt depends on the practical purpose. Man seems to approach the very same reality from different levels of action, and so the very same world may appear to him as a mechanistic total in which the individual elements are still clearly visible, or as a Gestalt where the parts themselves are fused, but some of the relationships between them are unconsciously emphasised.[136]

In the light of this outline of the practical sources of knowledge in society, and notwithstanding his appeal to field theory, and topology, it should come as no surprise, therefore, to find Kirk neglecting motivationally generated hodological structures, the varied, orientated, and social open-endedness of which is especially clear from the following:

> It is fundamentally this need for more adequate psychological definitions of direction that has interested Lewin in topological principles. ... The new body of concepts which Lewin has

developed on the basis of topology is really a sort of geometry particularly constructed to deal with the relationships of direction which characterize the different possible courses of action, or paths, within the life space. This geometry, although it is developed from topology, is topology no more, and needs a new name. Lewin has called it 'hodology' (from the Greek 'hodos', for way or path – hence hodology, the science of paths). Similarly, he speaks of 'hodological space' rather than 'topological space' when referring to psychological situations that must be understood to some extent in terms of direction or distance.[137]

Finally, if Kirk appears to divest Lewin's work of its chief, motivational thrust, then other contemporary geographers have done the same, and perhaps for not unlike, at root naturalistic reasons. Can Berry, for example, really be said to have done justice to Lewin's theory, and to Philbrick's 'areal functional organization' as an outgrowth of 'creative human choice',[138] when he proposed a synthesis of formal and functional regions using a general field theory of spatial behaviour?[139] And similarly, is not 'the stasis of diffusion theory' due, as Gregory imputes, to that same corrosive, naturalistic impulse that leads Hägerstrand to render human 'projects' as 'paths' in his 'time geography', and which the personal directedness of Lewin's actors in life space surely avoids?[140] Entrikin's[141] failure to mention Cassirer's considerable, indirect influence on geography, via Lewin, is the more understandable, therefore, since often, as Kirk's case shows, Lewin's social psychology has often been shorn of its hodological emphasis by modern geographers.

It can be said, in conclusion, that for certain geographers there is no doubt that Kirk's 1952 paper represents 'a watershed in British geography',[142] a harbinger of 'the behavioural revolution', and even an early forerunner of what is now termed 'humanistic geography'. For some, however, such a verdict is impossible. Baker, for example, judges Kirk's achievement as more modest:

> Kirk's paper did not inaugurate a new line of work – historians and historical geographers had long produced studies in perceptual geography – but it did make explicit a view of the world that was implicit in the writings of a number of others [and] ... did ... identify some of the psychological underpinnings of that work.[143]

Still others regard him as even less innovative. Eyles is typical:

> Kirk's work should perhaps be seen as the end of an era, that which emphasised society and environment in intricate

interrelation, rather than the start of a technicalist behavioural geography. That is not to say that Kirk's work is without contemporary relevance. It can form a focus for an emerging critical social geography.[144]

However perceptive, or not, these very diverse assessments are, they are all (like this one) inevitably, and some obviously, tinged with the highly dubious benefits of hindsight, and so must be considered with due reserve. Yet even when Kirk himself has reflected on his own earlier views, in the light of subsequent developments, and prospective advances, the picture has not become much clearer. Humanists, and Marxists, for instance, would doubtless draw comfort from Kirk's situating the current debate 'in the wake of the apparent collapse of over-ambitious scientific ideals'. But positivists too might draw equal succour from his conviction that 'within the spatial discipline of geography the concept of self-regulating transformation of structures is widely applicable, from geomorphology to the behavioural complexes of "political man"'.[145]

Ultimately, therefore, an understanding of Kirk's position can perhaps best be achieved by two means, used in combination: on the one hand, by acknowledging what he regards as the disappointingly small amount of 'perception' research done in either historical or political geography, and suggesting some possible reasons for this attributable to his own thought; and, on the other, by carefully noting what he has not declared, namely a thoroughgoing personal endorsement of 'humanist geography'.[146] Thus, some contemporary geographers might well be impressed with the foresight of Kirk's early attention to ideological matters (whether the misconceptions about one state held in another, or those enduring 'major perceptions' reflected in political behaviour and strategic thinking), or of his transactional approach to tactical warfare or frontier making. However, they might also complain that in the absence of his explicit discussion of ideology's implications for geography (political, or any other kind), or of any appropriate amendment to his diagrams to allow for transaction (political, or indeed of any human sort, economic, social, or cultural), these valuable insights remained hidden, only lately to re-emerge in the discipline, in bolder form, in the work of others, and Kirk's own more recent statements. Other geographers, of course, may feel that Kirk's emphasis on the role of perception, and particularly on the past designs, achievements, and mistakes of great men, presages a decidedly modern cultural and historical geography, configurative, holistic, and respectful of human agency.

On closer inspection, however, they too might soon be disappointed with an approach which makes seemingly very few philosophical concessions to human agency, and in practice is heedless of cultural or historical relativity in its compulsive pursuit of a naturalistic theory of geographical change. The result is geography very different from that in some of the post-war continental European traditions, where, as the case of Mannheim exemplifies, the cultural-historical sciences were already more powerfully developed, positivism was undermined earlier, and geographers pointed far sooner to the distinctiveness of human motivations, and purposive behaviour. Some may feel that in recent years Kirk has himself moved closer to this continental standpoint: his appeal to the work of Piaget and Lewin, both with their neo-Kantian undertones, would appear to indicate an infusion of idealism, and in addition a greater regard for social psychology, with its promise of a socially constituted, not merely aggregative representation of geographical space. Yet, Kirk's positivism here and elsewhere reasserts itself, and his desire for a newly invigorated environmentalism re-emerges; and chiefly responsible remains unquestionably his concept of geography as a unified earth science. Whether the related concepts of geographical, behavioural, and phenomenal environments adequately justify this view of geography seems doubtful, however, because their neglect by epistemologists may not just be a simple case of oversight (with the aid of hindsight), of a misperception of Kirk's original chief intention. For in the final analysis the main reason may well be that in proposing and elaborating these undoubtedly thoughtful concepts Kirk has nevertheless shown neither sufficient nor consistent awareness of those methodological and philosophical problems which a prescient, if sceptical Febvre knew human geography must sometime confront.

Notes

1 H. C. Darby, *The Theory and Practice of Geography: An Inaugural Lecture Delivered at Liverpool on 7 February 1946* (Liverpool University Press, Liverpool, 1946), p. 9.

2 'Human Geography. . . . Beware. For as long as it was a simple matter of a geography, as Vidal says, which concerned itself naturally and in a broad sense with mankind – no difficulties, no conflicts, no dangers [arose]. From the day on which a fully fledged autonomous science called human geography claims to have been founded, from the day on which mankind was thus formally placed centre stage, from that day, difficulties arose. And may I add philosophical as well as methodological ones.' (L. Febvre, 'Le problème de la géographie

humaine. A propos d'ouvrages récents', *Revue de Synthèse Historique*, vol. 35 (1923), pp. 114–15.)

3 W. Kirk, 'Historical geography and the concept of the behavioural environment', *Indian Geographical Journal*, Silver Jubilee Volume (1952), pp. 152–60; W. Kirk, 'Problems of geography', *Geography*, vol. 48 (1963), pp. 357–71.

4 See, for example, H. Capel, 'Percepción del medio y comportamiento geográfico', *Revista de Geografia*, vol. 7 (1973), p. 60; E. Jones, 'Social geography' in E. H. Brown (ed.), *Geography: Yesterday and Tomorrow* (Oxford University Press, Oxford, 1980), p. 254; A. Buttimer, *The Practice of Geography* (Longman, London and New York, 1983), p. 189; P. Claval, *Géographie humaine et économique contemporaine* (Presses Universitaires de France, Paris, 1984), p. 113; D. J. Walmsley and G. J. Lewis, *Human Geography: Behavioural Approaches* (Longman, London and New York, 1984), p. 7; R. G. Golledge and R. J. Stimson, *Analytical Behavioural Geography* (Croom Helm, London, 1987), p. 1.

5 H. Butterfield, *The Whig Interpretation of History* (Bell & Sons, London, 1931). 'It is part and parcel of the whig interpretation of history that it studies the past with reference to the present; and though there may be a sense in which this is unobjectionable if its implications are carefully considered, and there may be a sense in which it is inescapable, it has often been an obstruction to historical understanding because it has been taken to mean the study of the past with direct and perpetual reference to the present. ... Real historical understanding is not achieved by the subordination of the past to the present, but rather by our making the past our present and attempting to see life with the eyes of another century than our own' (pp. 11, 16).

6 A recent reminder of the need for such circumspection in the interpretation of the past has come from Darby, who writes: 'one must always remember the effect of hindsight upon the evaluation of contemporary conditions in past ages' (H. C. Darby, 'Historical geography in Britain, 1920–1980: continuity and change', *Transactions, Institute of British Geographers*, New Series, vol. 8 (1983), p. 424). A number of existing reviews of 'behavioural geography' do, it must be said, show the necessary caution, and appropriate wariness of putative historical 'reconstructions'. Thus, T. F. Saarinen (*Environmental Planning: Perception and Behavior* (Houghton Mifflin, Boston, 1976), p. 151) judges that 'geosophy did not catch on, nor did the term behavioral environment, proposed by William Kirk'; and, of the latter term, Brookfield likewise remarks that it 'has really achieved only retrospective notice' (H. C. Brookfield, 'On the environment as perceived', *Progress in Geography*, vol. 1 (1969), p. 57). Subsequently, Downs and Meyer have warned of the 'revisionist' dangers inherent in piecing together a developmental account of 'perceptual geography', of 'finding continuity where none was experienced and progress where none was felt. ... . These reconstructed family trees are of little value. They are post hoc

structures which do not adequately reflect the context of ideas within which the field emerged. It is more important to try to capture and reconstruct the context, the intellectual framework and atmosphere within which perceptual geography emerged' (R. M. Downs and J. T. Meyer, 'Geography and the mind. An exploration of perceptual geography', *American Behavioral Scientist*, vol. 22 (1978), pp. 59, 62). Of what some at least would regard as the embryonic stages of this emergence, moreover, Thrift argues: 'Whereas with many other subject areas it is possible to establish a theoretical bloodline that stretches back beyond the "quantitative revolution", it is hard to convincingly reconstitute the scattered set of papers that exist from before this point in time that can be labelled as "behavioural" as in any way representing a coherent behavioural geography' (N. Thrift, 'Behavioural geography', in N. Wrigley and R. J. Bennett (eds), *Quantitative Geography: A British View* (Routledge & Kegan Paul, London, (1981), p. 353). Furthermore, it is surely significant that Kirk himself has alluded to 'that often quoted but, I suspect, less frequently read, paper of mine in 1952. The concept of the behavioural environment first formulated therein has had a chequered history in the past 25 years of geographical thought and in the process perhaps sight has been lost of the context in which it was first defined' (W. Kirk, 'The road from Mandalay: towards a geographical philosophy', *Transactions, Institute of British Geographers*, New Series, vol. 3 (1978), p. 388).

7 As in an early study by D. Ley, *The Black Inner City as Frontier Outpost* (Association of American Geographers, Washington, DC, 1974), p. 7, and such markedly empiricist papers as those collected in R. G. Golledge and J. N. Rayner (eds), *Proximity and Preference. Problems in the Multi-dimensional Analysis of Large Data Sets* (University of Minnesota Press, Minneapolis, 1982).

8 Exemplified by L. J. Wood, 'Perception studies in geography', *Transactions, Institute of British Geographers*, vol. 50 (1970), pp. 129–42; M. Chisholm, *Research in Human Geography* (Heinemann, London, 1971), pp. 46–7; B. Goodey, 'Perception of the environment', Occasional Paper no. 17 (1971), Centre for Urban and Regional Studies, University of Birmingham; R. Geipel, 'La géographie de la perception en Allemagne Fédérale', *L'Espace Géographique*, vol. 3 (1978), pp. 195–8; P. Claval, 'L'évolution récente des recherches sur la perception', *Rivista Geografica Italiana*, vol. 87 (1980), pp. 6–24; J. R. Gold, *An Introduction to Behavioural Geography* (Oxford University Press, Oxford, 1980); A. Bailly, 'La géographie de la perception dans le monde francophone: une perspective historique', *Geographica Helvetica*, vol. 36 (1981), pp. 14–21; T. F. Saarinen, D. Seamon, and J. L. Sell (eds), 'Environmental perception and behavior: an inventory and prospect', Research Paper no. 209 (1984), Department of Geography, University of Chicago. Other prominent figures are, however, noticeably more tentative: for instance, the two convenors of an early, and influential, symposium

(K. R. Cox and R. G. Golledge (eds), 'Behavioral problems in geography: a symposium', Northwestern University Studies in Geography no. 17 (1969), Evanston, Ill.) have subsequently been careful to stipulate: 'It should be made clear that, in our view, the program of research endeavor envisaged by behavioral geography never involved the creation of a new branch of the discipline. Rather it was concerned with the elaboration of a distinctive approach to developing theory and solving problems in a wide variety of substantive areas in human geography' (K. R. Cox and R. G. Golledge (eds), *Behavioral Problems in Geography Revisited* (Methuen, New York and London, 1981), p.xxvi). Progress reports appearing periodically in the journal *Progress in Human Geography* encourage belief in the existence of a 'field' of study, but even their authors are a little equivocal on the matter: 'Working on the premise that behavioural and perceptual geography is not a rigidly constituted subdiscipline but rather a broad movement within geography . . .' (J. R. Gold and B. Goodey, 'Behavioural and perceptual geography', *Progress in Human Geography*, vol. 7 (1983), p. 578). The position has, of course, always been further complicated by the avowedly interdisciplinary nature of much research, as another review lately makes apparent (C. Spencer and M. Blades, 'Pattern and process: a review essay on the relationship between geography and environmental psychology', *Progress in Human Geography*, vol. 10 (1986), pp. 230–48).

9 Kirk, 'Problems of geography', p. 364.
10 Ibid., pp. 358–9.
11 Kirk, 'Historical geography and the concept of the behavioural environment', p. 158.
12 His 'experiences of ridge-to-ridge fighting against the Japs through the jungle monsoon of the Burma campaign, surrounded by death, injury and disease was grim, front-line war that few of us could even imagine' (J. G. Cruickshank, 'Professor William Kirk, O.B.E. An appreciation', *Graticule*, no. 29 (1986), p. 7). Not dissimilar, but more desperate, was the fate of C. A. Fisher as one of the prisoners of war used by the Japanese to construct the infamous Burma–Siam railway. The first part of Fisher's *Three Times a Guest: Recollections of Japan and the Japanese, 1942–1969* (Cassell, London, 1979) contains graphic descriptions of the terrible conditions of imprisonment and forcible 'coolie' labour suffered by British and other POWs on the railway; and his *South-East Asia: A Social, Economic and Political Geography*, 2nd edn (Methuen, London, 1966) opens with an ironic reference to the unique insight hard won from his enforced wartime 'participant observation': 'For in living for several years at a level little different from that of millions of Asian peasants, and making the best one could of an environment that often seemed to pose more problems than it afforded opportunities, I learned in some degree to look at South-east Asia from within rather than, as I had hitherto done, from without, and this surely is the beginning of geographical understanding' (p. vii).

That many 'returning warriors' owed their post-war professional interests to such unforeseen, and sometimes punishing, experiences of the humid tropics, and other regions abroad, is a point well made by B. H. Farmer, 'British geographers overseas, 1933–1983', *Transactions, Institute of British Geographers*, New Series, vol. 8 (1983), pp. 70–9 and W. G. V. Balchin, 'UK geographers in the Second World War. A report', *Geographical Journal*, vol. 153 (1987), pp. 159–80. See also the Preface to R. W. Steel and C. A. Fisher (eds), *Geographical Essays on British Tropical Lands* (Philip, London and Liverpool, 1956).

13 W. Kirk and H. Godwin, 'A late glacial site at Loch Droma, Ross and Cromarty', *Transactions of the Royal Society of Edinburgh*, vol. 65 (1963), pp. 225–49; W. Kirk, R. J. Rice, and F. M. Synge, 'Deglaciation and vertical displacement of shorelines in Wester and Easter Ross', *Transactions, Institute of British Geographers*, vol. 39 (1966), pp. 65–78; W. Kirk, 'The lower Ythan in prehistoric times', in J. Goldman (ed), *A History of the Burgh and Parish of Ellon* (W. and W. Lindsay, Aberdeen, 1958); and W. Kirk, 'Prehistoric sites at the sands of Forvie, Aberdeenshire. A preliminary examination', *Aberdeen University Review*, vol. 35 (1953), pp. 150–71.

14 W. Kirk, 'Review of D. R. Stoddart, *On Geography*', *Journal of Historical Geography*, vol. 13 (1987), p. 108.

15 Kirk, 'Problems of geography', pp. 364–5. For instance, Geertz insists that culture is not a characteristic of humanity superadded to human biological structure, but instead that biological and cultural evolution proceeded in step one with the other and that the development of human culture contributed greatly to the direction of human biological evolution. See C. Geertz, 'The impact of the concept of culture on the concept of man', in Y. A. Cohen (ed.), *Man in Adaptation. The Cultural Present* (Aldine, Chicago, 1968), pp. 16–29.

16 Kirk, 'Historical geography and the concept of the behavioural environment', p. 159.

17 S. W. Wooldridge, 'On taking the "ge-" out of geography', *Geography*, vol. 34 (1949), p. 17.

18 'Now this movement does not take place in the physical environment of this page but in the mind dependent on the stimuli of light passing through the eyes' (Kirk, 'Historical geography and the concept of the behavioural environment', p. 159).

19 Kirk, 'Review of D. R. Stoddart, *On Geography*', p. 109.

20 As when he asserts: 'Some peoples are drawn to the sea by the physical nature and economic potential of its coastlands, others are driven to it when the land environment no longer provides at a given technological level the needs of a growing population' (W. Kirk, 'Indian Ocean community', *Scottish Geographical Magazine*, vol. 67 (1951), p. 164). The basis for this, and kindred claims, is clear from the following statement: 'I have always believed that the behavioural environment/ phenomenal environment model subsumed the traditional stimulus–response model that had dominated earlier geographical thought and

led to the separation of "man" and "environment"' (Kirk, 'The road from Mandalay', p. 389).

21 Comprehending the prehistoric geography of Scotland depends, he maintains, on an appreciation of the fact that the 'first farmers were confronted then by a great forest, as yet virtually untouched by man, and consequently the springs of their behavioural patterns can be sought in the varied potentials of different forest tracts' (W. Kirk, 'The primary agricultural colonisation of Scotland', *Scottish Geographical Magazine*, vol. 73 (1957), p. 77). What is true of Scotland he evidently feels applies to Norway (see W. Kirk and F. M. Synge, 'Farms of Verdal, Norway', *Scottish Geographical Magazine*, vol. 70 (1954), pp. 106–23, esp. p. 122), and indeed Western Europe in its entirety, as the following statement shows: 'In view of the fact that for the major part of our history we Western Europeans were a forest folk, and our culture still bears a considerable imprint of that environmental experience, this attempt to tell our story in woodland terms must be commended – and Ireland offers a classic case of the destructive impact of man on a forest environment' (W. Kirk, 'Review of F. Mitchell, *The Irish Landscape*', *Progress in Human Geography*, vol. 1 (1977), pp. 149–50). In similar vein, he concludes his treatment of the prehistory and early history of the Indian subcontinent with observations about the role of the forest as a preserver and transformer of cultural traits: thus, of the latter innovative role, he remarks: 'Ideas and cultural systems from the west – whether one is discussing architectural traits, town planning, metallurgy, or agricultural techniques – rarely remain unmodified for long. They are absorbed, reorganised and transformed in India to meet the needs of forest environments, and the cultural package ultimately transmitted eastward to South-east Asia is often hardly recognizable as one received earlier from the west. In this respect, India has functioned as a great selective filter in cultural diffusion' (W. Kirk, 'The role of India in the diffusion of early cultures', *Geographical Journal*, vol. 141 (1975), p. 33).

22 K. H. Paffen, 'Stellung und Bedeutung der physischen Anthropogeographie', *Erdkunde*, vol. 13 (1959), pp. 354–72.

23 J. H. Bird, 'Geography in three worlds: how Popper's system can help elucidate dichotomies and changes in the discipline', *Professional Geographer*, vol. 37 (1985), p. 403; J. H. Bird, 'Methodological implications for geography from the philosophy of K. R. Popper', *Scottish Geographical Magazine*, vol. 91 (1975), pp. 153–63.

24 Kirk, 'Historical geography and the concept of the behavioural environment', p. 159.

25 'The villain of the piece in much of the unclarified thinking on this matter is the overworked word "environment". It can signify so obviously almost everything as to come to mean almost nothing' (Wooldridge, 'On taking the "ge-" out of geography', p. 10).

26 The epistemological import of Kirk's schema is, however, recognized by H. Uhlig, 'System der Geographie', *Westermann Lexikon der Geographie*, vol. 4 (1970), p. 495.

27 Kirk, 'The road from Mandalay'.
28 W. Kirk, 'The high road from Manchester to Aberdeen in the immediate post-war years', in W. Ritchie, J. C. Stone, and A. S. Mather, *Essays for R. E. H. Mellor* (Department of Geography, University of Aberdeen, 1986), p. 5. For a geographer's account of the transformation of British universities in the post-war period see W. G. V. Balchin, 'University expansion in Great Britain', *New Scientist*, vol. 5 (1959), pp. 582–5.
29 C. A. Fisher, 'Editorial. The third generation', *Geographical Studies*, vol. 1 (1954), p. 1.
30 Kirk, 'Historical geography and the concept of the behavioural environment', p. 158.
31 R. Hartshorne, *Perspective on the Nature of Geography* (Murray, London, 1960), p. 58.
2 O. H. K. Spate, 'How determined is possibilism?', *Geographical Studies*, vol. 4 (1957), p. 11.
33 It is worth noting here an earlier remark of Wooldridge (and Linton) who wrote of 'the risk of replacing the crudities of "Determinism" by the banalities of a formless "Possibilism"' (S. W. Wooldridge and D. L. Linton, 'Structure, surface and drainage in south-east England', Institute of British Geographers Publication no. 10 (1939), Philip, London, p. 124).
34 H. Sprout and M. Sprout, *Man–Milieu Relationship Hypotheses in the Context of International Politics* (Center of International Studies, Princeton University, Princeton, NJ, 1956); H. Sprout and M. Sprout, *The Ecological Perspective on Human Affairs with Special Reference to International Politics* (Princeton University Press, Princeton, NJ, 1965).
35 O. H. K. Spate, 'Environmentalism', *International Encyclopaedia of the Social Sciences*, vol. 5 (1969), p. 94.
36 M. Billinge, 'In search of negativism: phenomenology and historical geography', *Journal of Historical Geography*, vol. 3 (1977), p. 59; and R. J. Johnston, *Geography and Geographers. Anglo-American Human Geography since 1945*, 3rd edn (Edward Arnold, London, 1987), p. 148. For an assessment which interprets Kirk's thinking as neither positivist nor idealist, see E. M. W. Gibson, 'Realism', in M. E. Harvey and B. P. Holly (eds), *Themes in Geographic Thought* (Croom Helm, London, 1981), p. 155.
37 An early affirmation of his adherence to positivism can be found in Kirk, 'Historical geography and the concept of the behavioural environment', p. 153.
38 W. Kirk, 'Damodar valley – *valles opima*', *Geographical Review*, vol. 40 (1950), p. 420.
39 'The environmental-relationship viewpoint remained environmentalism even if the reciprocal character of relationships be stressed, and even if the "possibilist" contention (that nature offers man a wide range of choice) be insisted upon' (G. R. Lewthwaite, 'The nature of environmentalism', Proceedings of the Second New Zealand Geography Conference, Christchurch, August 1958, p. 6).

40 An account of the Umweltlehre can be found in J. J. von Uexküll, *Umwelt und Innenwelt der Tiere* (J. Springer, Berlin, 1909); M. Schwind, 'Die Umweltlehre Jakob von Uexkülls in ihrer Bedeutung für die Kulturgeographie', Verhandlungen des 28 Deutschen Geographentag, Frankfurt am Main 1951, pp. 291–5; M. Schwind, 'Kulturlandschaft als objektivierter Geist', *Deutsche Geographische Blätter*, vol. 46 (1951), pp. 4–28; M. Schwind, 'Die politische, social und kulturelle Aussage der Landschaft', *Gesellschaft, Staat, Erziehung*, vol. 8 (1963), pp. 351–9; and M. Schwind, *Kulturlandschaft als geformter Geist. Drei Aufsätze über die Aufgaben der Kulturgeographie* (Wissenschaftliche Buchgesellschaft, Darmstadt, 1964).

41 M. Schwind, *Das Japanische Inselreich. Band 2: Kulturlandschaft, Wirtschaftsgrossmacht auf engem Raum* (de Gruyter, Berlin and New York, 1981), pp. 218–24.

42 'the period of disruption' (A. Meynier, *Histoire de la pensée géographique en France* (Presses Universitaires de France, Paris, 1969), p. 117).

43 'L'éclatement des genres de vie anciens par suite de la naissance de formes nouvelles de production ne laisse plus à la place de l'ancienne unité économique et sociale rurale qu'une poussière de genres de vie professionels'/'The shattering of old established *genres de vie* as a consequence of the emergence of new forms of production permits henceforth instead of the erstwhile social and economic unity of rural life only a profusion of professional *genres de vie*' (P. George, *Introduction à l'étude géographique de la population du monde* (Presses Universitaires de France, Paris, 1951), p. 72).

44 P. George, 'Sur une nouvelle présentation du déterminisme en géographie humaine', *Annales de Géographie*, vol. 61 (1952), pp. 280–4.

45 Kirk, 'Problems of geography', p. 360.

46 '*Genre de vie* is especially a descriptive device, reasoned description to be sure, but in which explanation still only accompanies and sustains the description without being able to disengage itself from it and still less precede it' (J. Gottmann, 'De la méthode d'analyse en géographie humaine', reprinted in *Essais sur l'aménagement de l'espace habité* (Mouton, Paris, 1966), p. 81).

47 'the most quintessentially human aspect: the psychological factor' (Gottmann, 'De la méthode d'analyse en géographie humaine', p. 89).

48 'In human geography, the analytical method must always take account of the spiritual factor, investigate the psychological ferment, appreciate its strength. We must steer clear somewhat of the logic of the experimenters, especially the natural scientists: Claude Bernard could say: "The fact judges the idea", for we always desire factual evidence. But it is the idea which gives rise to the fact, doubtless indirectly and by opening the door to many surprises, but psychological activity (nonetheless) is fundamental to the dynamism of human geography (and we encapsulate in this sense, in "human", the

economic as well as the political and social). A method of analysis, to be scientific in our discipline, ought then to repudiate a geographical materialism too simplistically related to the facts and allow for the spiritual forces which can offset pressures the energies of which are drawn from other sources' (Gottmann, 'De la méthode d'analyse en géographie humaine', pp. 90–1).

49 J. Gottmann, *Megalopolis, the Urbanized North-Eastern Seaboard of the United States* (Twentieth Century Fund, New York, 1961), p. 23.

50 J. Gottmann, 'The political partitioning of our world: an attempt at analysis', in W. A. D. Jackson and M. S. Samuels (eds), *Politics and Geographic Relationships*, 2nd edn (Prentice-Hall, Englewood Cliffs, NJ, 1971), p. 272.

51 G. H. T. Kimble, 'The inadequacy of the regional concept', in L. D. Stamp and S. W. Wooldridge (eds), *London Essays in Geography* (Longman, London, 1951), pp. 151–74.

52 Kirk, 'Historical geography and the concept of the behavioural environment', p. 156.

53 Ibid., p. 158.

54 Ibid., p. 155.

55 Ibid., p. 154.

56 Kirk, 'The primary agricultural colonisation of Scotland', p. 84.

57 W. Kirk (with N. C. Mitchel), 'South-east Asia: a study of cores and peripheries in development processes' (mimeograph, no date), pp. 40, 1.

58 W. Kirk, 'A genetic model of frontier making', Presidential Address, Section E, British Association for the Advancement of Science, Belfast Meeting, 24–28 August 1987 (mimeograph), p. 14.

59 W. Kirk, 'Cores and peripheries: the problems of regional inequality in the development of southern Asia', *Geography*, vol. 66 (1981), p. 200.

60 Ibid.

61 Kirk, 'A genetic model of frontier making', p. 15.

62 Typical is his genetic model of frontier making, which is cyclical, and has clearly marked stages, and phases: 'with the collision of two or more politico-territorial systems a primary stage is reached in which the major contestants leave a vaguely defined frontier zone between them, a secondary stage in which the frontier zone acquires internal political form expressed in the creation of buffer states, and a tertiary stage in which the zone collapses by peripheral political pressures and alignment of buffer states to a boundary line capable of precise definition' (W. Kirk, 'The Sino-Indian marchlands', in W. G. East, O. H. K. Spate, and C. A. Fisher (eds), *The Changing Map of Asia. A Political Geography* (Methuen, London, 1971), p. 193). The inevitability of the changes described is clear from Kirk's outline of historical events in one of the phases: 'Those in front cry forward and those behind cry halt, but the mechanism of frontier turbulence ensures forward impetus' (Kirk, 'A genetic model of frontier making', p. 11).

Reflection

63 Kirk, 'A genetic model of frontier making', p. 4.
64 W. Kirk, 'Prehistoric Scotland: the regional dimension', in C. M. Clapperton (ed.), *Scotland: A New Study* (David and Charles, London, 1983), p. 112.
65 W. Kirk, *Geographical Pivots of History: An Inaugural Lecture Delivered in the University of Leicester, 24 November 1964* (Leicester University Press, Leicester, 1965), p. 24.
66 Kirk, 'Indian Ocean community'.
67 O. H. K. Spate, *The Spanish Lake* (Croom Helm, London, 1979); and *Monopolists and Freebooters* (Croom Helm, London and Canberra, 1983).
68 O. H. K. Spate, 'The Pacific as an artefact', in N. Gunson (ed.), *The Changing Pacific. Essays in Honour of H. E. Maude* (Oxford University Press, Melbourne, 1978), pp. 32–3.
69 Ibid., p. 44.
70 O. H. K. Spate, 'Prolegomena to a history of the Pacific', *Geographia Polonica*, vol. 36 (1977), p. 218.
71 Opinions expressed by the archaeologist Renfrew come to mind here, for he has advanced such arguments as: 'The very nature of the data base establishes an overlap between archaeology and physical geography, and to a large extent conditions archaeology (and perhaps more of geography than is currently accepted) to be a fundamentally empiricist or 'positivist' discipline rather than an idealist one .... The empiricist/positivist geographer and the archaeologist thus often deal with operational rather than cognized data – with things as they appear to be to an outsider, rather than as they are known to be to an insider. This is indeed, quite validly, part of the idealist criticism of both subjects' (C. Renfrew, 'Space, time and man', *Transactions, Institute of British Geographers*, New Series, vol. 6 (1981), pp. 257, 260).
72 W. Kirk, 'North-East Scotland', in J. B. Mitchell (ed.), *Great Britain. Geographical Essays* (Cambridge University Press, Cambridge, 1962), pp. 511, 518. For a contrasting, decidedly 'insider' portrayal of this region, and its places, see the essay by a native 'in-dweller' (his term), Alexander Fenton, 'Aspects of the North-East personality', in *The Shape of the Past. 1: Essays in Scottish Ethnology* (John Donald, Edinburgh, 1985), pp. 56–67.
73 O. H. K. Spate, 'Quantity and quality in geography', *Annals of the Association of American Geographers*, vol. 50 (1960), p. 384.
74 W. Kirk, 'The cotton and jute industries of India: a study in concentration and dispersal', *Scottish Geographical Magazine*, vol. 72 (1956), p. 39.
75 G. W. S. Robinson, 'The geographical region: form and function', *Scottish Geographical Magazine*, vol. 69 (1953), p. 50.
76 'too mechanistic' ... 'emerging cultural, historically conditioned particularities' (G. Pfeifer, 'Zur Funktion des Landschaftsbegriffes in der deutschen Landwirtschaftsgeographie', *Studium Generale*, vol. 11 (1958), p. 409).
77 'the driving force in the realm of psychical causality is human resolve to a common purposive influence upon Nature or upon the cultural

landscape moulded by earlier generations' (H. Lautensach, 'Otto Schlüters Bedeutung für die methodische Entwicklung der Geographie', *Petermanns Geographische Mitteilungen*, vol. 96 (1952), p. 223).

78 'Created autonomously by the mind of mankind they [social geographical forces] accordingly do not know causalities but only motivations. What counts as far as they are concerned are the rules, the principles and the laws for the acts and effects of judgement and intention within individuals, groups and communities which have been established by sociology. If it wishes to raise itself to scientific status, geography must take cognisance of these fundamentals and introduce them to the play of forces in the landscape' (H. Bobek, 'Stellung und Bedeutung der Sozialgeographie', *Erdkunde*, vol. 2 (1948), p. 122).

79 J. Labasse, *L'Organisation de l'espace. Eléments de géographie volontaire* (Hermann, Paris, 1966); E. A. Ackerman, 'A view of terrestrial space', *Science*, vol. 157 (1967), pp. 1031–2; and E. A. Ackerman, 'Geography as a fundamental research discipline', Research Paper no. 53 (1958), Department of Geography, University of Chicago, p. 36: 'If geography is to be interested in process it cannot afford to ignore basic techniques already developed for behavioural study.'

80 W. Kirk (with F. W. Boal and W. B. Whalley), 'Questions of integration in geography', Video-tape G35 recorded at Queen's University, Belfast, January 1984, and released by Invitation to Dialogue Project, Department of Geography, University of Lund, Sweden.

81 Golledge, for instance, cites as one of the 'set of myths building up around behavioral research in geography' the supposition that 'a behavioral approach is just as valid for studying the activities of inanimate systems as it is for examining sensate behavior' (R. G. Golledge, 'Misconceptions, misinterpretations, and misrepresentations of behavioral approaches in human geography', *Environment and Planning A*, vol. 13 (1981), p. 1325).

82 Kirk, 'Problems of geography', p. 368.

83 Hartshorne, *Perspective on the Nature of Geography*, pp. 152–3.

84 E. Jones, *A Social Geography of Belfast* (Oxford University Press, London, 1960), p. 47.

85 M. S. Samuels, 'The biography of landscape: cause and culpability', in D. W. Meinig (ed.), *The Interpretation of Ordinary Landscapes: Geographical Essays* (Oxford University Press, New York and Oxford, 1979), p. 84.

86 Kirk, *Geographical Pivots of History*, pp. 14, 12.

87 Kirk, 'Problems of geography', p. 368. Elsewhere he speaks of 'cultural levels' ('Questions of integration in geography').

88 Kirk (with Mitchel), 'South-east Asia', p. 2.

89 W. Kirk (with N. C. Mitchel), 'Development problems in post war restructuring of core periphery systems' (mimeograph, no date), p. 41.

90 Ibid., p. 62.

91 Kirk, 'The road from Mandalay', p. 388.

92 W. Kirk, 'Devolution in the UK: revolt of the Celts?', *Geographical Magazine*, vol. 50 (1978), p. 756.
93 Kirk (with Mitchel), 'South-east Asia', p. 3.
94 Kirk, 'Cores and peripheries', p. 200.
95 Kirk, 'A genetic model of frontier making', p. 4.
96 Kirk, 'The road from Mandalay', p. 389.
97 A student of international affairs, Joseph Frankel participated, for instance, in the seminar convened at the University of Aberdeen during the Sprouts' visit (Kirk, 'The road from Mandalay', p. 393). And having referred at the start of his book, *The Making of Foreign Policy* (Oxford University Press, London, 1963), p. 4, to 'some social scientists (including a few geographers) who may be called "cognitive behaviourists"', he proceeds to outline a model in which 'following Professor and Mrs Sprout, a distinction is ... made between psychological and operational environments, the former as apperceived by the decision makers, the latter as could be apperceived by an "omniscient observer".' As examples of other political works by Frankel see *International Relations* (Oxford University Press, London, 1964), *International Politics* (Allen Lane, London, 1969), and *International Relations in a Changing World* (Oxford University Press, Oxford, 1979).
98 J. W. House, *Frontier on the Rio Grande. A Political Geography of Development and Social Deprivation* (Clarendon Press, Oxford, 1982); J. W. House, 'The frontier zone: a conceptual problem for policy makers', *International Political Science Review*, vol. 1 (1980), pp. 456–77; J. W. House, 'Frontier studies: an applied approach', in P. J. Taylor and A. Burnett (eds), *Political Studies from Spatial Perspectives* (John Wiley, Chichester, 1981), pp. 291–312.
99 J. K. Wright, 'Training for research in political geography', *Annals of the Association of American Geographers*, vol. 34 (1944), pp. 190–1. Whereas Kirk's work is referred to in both the introductory section on 'Behaviour' in R. E. Kasperson and J. V. Minghi (eds), *The Structure of Political Geography* (University of London Press, London, 1970), pp. 299–318, and in the opening pages of the more recent R. Muir and R. Paddison, *Politics, Geography and Behaviour* (Methuen, London, 1981), neither makes reference to this paper by Wright. Interestingly, however, Lowenthal, an admirer of Wright, does preface his discussion of inter-island perceptions in the ill-fated West Indies Federation with the general observation: 'Men behave in accordance with their own images of themselves, of others, and their surroundings' (D. Lowenthal, 'The West Indies chooses a capital', *Geographical Review*, vol. 48 (1958), p. 337).
100 Kirk, 'A genetic model of frontier making', p. 12.
101 W. Kirk, 'The inner Asian frontier of India', *Transactions, Institute of British Geographers*, vol. 31 (1962), p. 134.
102 'The recent crisis of Democracy and Liberalism should bring home to those countries which still enjoy freedom some of the deficiencies of their system in the changed conditions of the world. ... A general

psychological break-down can only be prevented if we are quick
enough to realize the nature of the new situation, and to re-define the
aims and means of democratic education accordingly' (K. Mannheim,
Diagnosis of our Time (Routledge & Kegan Paul, London, 1943),
pp. 73–4).

103 At the end of his essay on the concept of Weltanschauung, for
example, Mannheim avers: 'The mechanistic method by which the
material is broken up into atomic constituents no longer appears
fruitful when it is applied to higher-level phenomena of meaning. In
the realm of the mental we cannot understand the whole from the
parts; on the contrary, we can only understand the parts from the
whole' (K. Mannheim, *Essays on the Sociology of Knowledge*
(Routledge & Kegan Paul, London, 1952), p. 82).

104 Kirk, *Geographical Pivots of History*, pp. 23–4.

105 It is interesting to read that two other geographers also find utopian
writers 'oblivious to, or impatient with, concerns that animate
geographers', utopia invariably being conceived as 'ahistorical,
aspatial within its boundaries, in time-space equilibrium, and with
resource problems either willed away or resolved through
technology' (P. W. Porter and F. E. Lukermann, 'The geography of
utopia', in D. Lowenthal and M. J. Bowden (eds), *Geographies of the
Mind* (Oxford University Press, New York, 1976), p. 216).

106 Kirk, 'Cores and peripheries', p. 200.

107 L. A. Coser, *Masters of Sociological Thought*, 2nd edn (Harcourt,
Brace, Jovanovich, San Diego, 1977), p. 435.

108 W. Kirk, 'The geographical significance of Vitruvius' *De
Architectura*', *Scottish Geographical Magazine*, vol. 69 (1953),
pp. 1–10; W. Kirk, 'Town and country planning in ancient India
according to Kautilya's *Arthasastra*', *Scottish Geographical
Magazine*, vol. 94 (1978), pp. 67–74.

109 In fact Kirk does make at least some concessions to such criticism of
historical reconstructions based on 'plan images' like that of Kautilya
where he writes: 'In view of the noisesome Indian cities of medieval
and later times it is difficult to imagine the orderly, disciplined cities
depicted in the *Arthasastra* and outlined in even greater detail in the
Manasara Silpasastra – the most comprehensive statement of the
early science of Indian town planning. . . . But even if no city was ever
precisely constructed according to the rules advocated in the text, nor
any kingdom organised entirely on the model recommended, the
ideals encoded in the *Arthasastra* provide considerable insight into
the perceptions and behavioural environments of the town and
country planners in early India, and facilitate our understanding of
the processes and relicts of Indian urbanisation dating from a vital
formative period' (Kirk, 'Town and country planning in ancient India
according to Kautilya's *Arthasastra*', pp. 67, 74, 75).

110 K. Mannheim, *Ideology and Utopia* (Routledge & Kegan Paul,
London, 1936), p. 3.

111 Ibid., p. 44.

112 The tentativeness here is deliberate, as Kirk himself seems unsure on the subject, and what he does say does not always seem consistent. Thus, on the one hand, he can acknowledge as a socially engaged but disinterested academic that 'the criterion of desirability' enters into the calculus of decision makers, and that 'in play-off situations . . . the framework of ranking is to a large measure socially or culturally determined', but, on the other hand, he can write of 'the just need to develop areas to the optimum condition possible and commensurate with their unequal resource bases', heedless now apparently of the different cultural appraisals, and various social objectives, that make identification of a single optimum an impossibility. Whereas the first position would also be that of Mannheim, the second, probably more representative of Kirk's standpoint, is in accordance with the notion of value-free social science (Kirk, 'The road from Mandalay', pp. 391, 392, 394).

113 'If we are to survive at all, and breed true, then some structures must be held steady, usually either as superstructures or infrastructures. We cannot take decisions on everything or we return to chaos. But the establishment of those structures takes time and leads us once more into the realms of philosophy' (Kirk, 'The road from Mandalay', p. 392).

114 'Though Gestalt psychology developed originally in the ambience of phenomenology, it retained only the phenomenologists' emphasis upon interaction between subject and object and was resolutely naturalistic in its orientation. Köhler had, after all, been trained as a physicist, and it was physics which suggested the notion, fundamental to him and other psychologists, of the "field". As we shall see, stimulating as the introduction of field models was initially, the dominant role assigned to them had some ill effects' (J. Piaget, *Structuralism* (Routledge & Kegan Paul, London, 1971). p. 54).

115 Kirk, 'Historical geography and the concept of the behavioural environment', p. 158.

116 Kirk, 'Problems of geography', p. 366.

117 W. Köhler, *Gestalt Psychology* (Bell & Sons, London, 1930), pp. 63, 133.

118 Ibid., pp. 115–16.

119 M. E. Eliot Hurst, *A Geography of Economic Behavior. An Introduction* (Duxbury Press, North Scituate, Mass., 1972), p. 45.

120 'The perceptual phenomena of special concern to all the Gestalt psychologists . . . were experiential counterparts of corresponding physical phenomena: perceptual Gestalten were isomorphic with physical Gestalten' (W. N. Dember and M. Bagwell, 'A history of perception', in G. A. Kimble and K. Schlesinger (eds), *Topics in the History of Psychology*, vol. 1 (Lawrence Erlbaum Associates, Hillsdale, NJ and London, 1985), pp. 271–2. The substance of this point receives at least some recognition from Kirk where he writes: 'In fact in the 1963 paper I did try to demonstrate how distortion of the image of the external world was related to processes of selection

of information flows by the intervention of culture/values filters and by the impact of learning on successive cognitive transformations of such processes. I accept, however, that there is a considerable amount of work still to be done on the topology of perceptual structures that is of geographical significance and *is not entirely covered in gestalt research*' (my emphasis) (Kirk, 'The road from Mandalay', p. 389).

121 K. Koffka, *Principles of Gestalt Psychology* (Kegan Paul, London, 1935), p. 648.

122 D. Katz, *Gestalt Psychology. Its Nature and Significance* (Methuen, London, 1951), pp. 163–4.

123 In the light of a statement earlier about the strong case for 'natural geography' put by Wooldridge, this assessment might seem surprising, but is surely warranted by the following: 'The physical landscape, including the vegetation cover, is the record of *processes* and the whole of the evidence for its evolution is contained in the landscape itself. The cultural landscape is the record of *actions* and we cannot interpret the actions without, in some sense, getting into the minds and reconstructing the thoughts of the men who so acted' (S. W. Wooldridge and F. Goldring, *The Weald* (Collins, London, 1953), p. 165). These sentences are from the chapter on pre- and early history, and the remarkable similarity between the sense of the latter sentence, and Collingwoodian ideas about history, is not happenstance, for the book's bibliography (on p. 261) includes R. G. Collingwood and J. N. L. Myres, *Roman Britain and the English Settlement* (Claredon Press, Oxford, 1937). For a recent application of Collingwood's approach in historical geography see L. Guelke, *Historical Understanding in Geography: An Idealist Approach* (Cambridge University Press, Cambridge, 1982).

124 Kirk, 'The road from Mandalay', p. 390.

125 'This (Gestalt) theory is the prototype of a structuralism without genesis, the structures being permanent and independent of development. I know very well that Gestalt theory has furnished concepts and interpretations of development itself, such as Koffka's [1928] excellent book on mental growth. Nevertheless, for him development is determined entirely by maturation, i.e. by a preformation which itself obeys Gestalt laws. Genesis remains secondary to the fundamental preformist perspective' (J. Piaget, *Six Psychological Studies* (University of London Press, London, 1968), p. 146).

126 J. Piaget, *Biology and Knowledge* (Edinburgh University Press, Edinburgh, 1971), p. 248.

127 J. Piaget, *The Growth of Logical Thinking* (Routledge & Kegan Paul, London, 1966), p. 243.

128 Kirk, 'The road from Mandalay', p. 393.

129 K. Mannheim, *Essays on Sociology and Social Psychology* (Routledge & Kegan Paul, London, 1953), p. 188.

130 Kirk, *Geographical Pivots of History*, p. 23.

131 A. J. Marrow, *The Practical Theorist. The Life and Work of Kurt Lewin* (Basic Books, New York, 1969), p. 9.
132 P. Fraisse and J. Piaget (eds), *Experimental Psychology. Its Scope and Method. Vol. 1: History and Method* (Routledge & Kegan Paul, London, 1968), p. 71.
133 Relevant here are some of the arguments made in J. C. Alexander, 'The "individualist" dilemma in phenomenology and interactionism', in S. N. Eisenstadt and H. J. Helle (eds), *Macro-sociological Theory. Perspectives on Sociological Theory*, vol. 1 (Sage, London, 1985), pp. 25–57.
134 Kirk, 'The road from Mandalay', p. 387.
135 A. Giddens, *New Rules of Sociological Method* (Hutchinson, London, 1976), p. 21.
136 Mannheim, *Essays on Sociology and Social Psychology*, p. 180.
137 R. W. Leeper, 'Lewin's topological and vector psychology', University of Oregon Monographs, Studies in Psychology no. 1 (1943), Eugene, Oregon, p. 37.
138 Though Philbrick's paper, like Robinson's mentioned above, includes the ambiguous adjective 'functional', its qualification of 'regional', its references to 'organization', and especially to human creativity, are all redemptive (A. K. Philbrick, 'Principles of areal functional organization in regional human geography', *Economic Geography*, vol. 33 (1957), p. 300).
139 B. J. L. Berry, 'A synthesis of formal and functional regions using a general field theory of spatial behavior', in B. J. L. Berry and D. F. Marble (eds), *Spatial Analysis* (Prentice-Hall, Englewood Cliffs, NJ, 1968), pp. 419–28.
140 D. Gregory, 'Suspended animation: the stasis of diffusion theory', in D. Gregory and J. Urry (eds), *Social Relations and Spatial Structures* (Macmillan, London, 1985), p. 325. It is perhaps not without significance that in this essay on Hägerstrand's theory, Gregory does not refer to Lewin, even though Hägerstrand himself, in his interview with Buttimer, does allude to Lewin's influence. See Buttimer, *The Practice of Geography*, p. 248.
141 J. N. Entrikin, 'Geography's spatial perspective and the philosophy of Ernst Cassirer', *Canadian Geographer*, vol. 21 (1977), pp. 209–22.
142 J. W. Watson, 'The soul of geography', *Transactions, Institute of British Geographers*, New Series, vol. 8 (1983), p. 397.
143 A. R. H. Baker, 'Historical geography in Britain', in A. R. H. Baker (ed.), *Progress in Historical Geography* (David and Charles, Newton Abbot, 1972), p. 94.
144 J. Eyles, 'Britain', in J. Eyles (ed.), *Social Geography in International Perspective* (Croom Helm, London and Sydney, 1986), p. 73.
145 Kirk, 'The road from Mandalay', pp. 382, 391.
146 Despite the attribution of this position to him by, for example, R. J. Johnston. See R. J. Johnston, *On Human Geography* (Basil Blackwell, Oxford, 1986), p. 55; and R. J. Johnston, *Philosophy and Human Geography*, 2nd edn (Edward Arnold, London, 1986), p. 68.

Chapter four

Environment, behaviour, and thought

Yi-Fu Tuan

'One person may enter a drugstore to buy medicine for a friend, another may enter to buy poison for an enemy' (Roger Barker).[1] If we observe only the behaviour, nothing perhaps distinguishes the one from the other. They both make the gestures appropriate to a drugstore, even though the worlds in their heads are radically different. A problem in the study of the relationship between environment and behaviour is that we do not know what is going on in people's heads. What goes on there may have little or nothing to do with the environment of the time and yet have much to do with subsequent action. This phenomenon is unique to human beings. It is, of course, also a commonplace of human experience. Ignoring it, as geographers and environmental psychologists have tended to do, gives their research – for all the wealth of empirical data – an air of unreality.

Human beings sometimes think. What is thinking? Difficult as it is to answer this question neurophysiologically, we all know from experience what it is like. Thinking is a special kind of activity; and as with the proper exercising of any organ, when we think well we have the pleasant sensation of being alive and in touch with the real. On the other hand, thinking is also de-sensing: it dematerializes the world around us. The ordinary person's suspicion of the thinker is captured by the expression, 'He has taken leave of his senses.' He is out of touch with his immediate environment. Hannah Arendt writes: 'While thinking, I am not where I actually am; I am surrounded not by sense-objects but by images that are invisible to everybody else.' Thus thinking produces a paradoxical effect: the near is far away (that is, out of mind) and the far away is brought near.[2]

Language and story-telling

All human beings are thinkers if only because we are all users of language. In using language, we create images and worlds independent of our immediate environment. If speaking and listening are behaviour, then we have to say that behaviours uniquely important to human beings correlate but poorly with physical setting. This becomes obvious if by behaviour we mean not only movements of the larynx and lips, or eyebrows knitted in concentration, but also the emotions and ideas that accompany such physiological displacements. In contrast to cries of alarm, murmurs of contentment, mating calls, and other vocal sounds in the non-human world, human speech transcends the stimulus–response model. Under that model, one vocal ejaculation will stimulate another, or bring about some kind of bodily movement. Monkeys demonstrate this behaviour in the rainforest. They chatter without break, and so the total volume of sound swells to an ear-splitting loudness, not unlike what happens among people in a cocktail party. Monkeys rarely listen. No one is willing to withhold response and be momentarily silent so that complex mental events have a chance to emerge. Peter Stephenson once suggested provocatively that without this capacity for silence – for withholding vocal action – language could not have evolved.[3] An important part of being human is this ability not to respond to a stimulus in an observably behavioural way. What I have described is quite an ordinary human experience that is already perfectly illustrated in early childhood when, to the sounds coming out of the parent's mouth, the child refrains from vocalizing him- or herself. Rather he listens: he is listening to a fairy-tale and, while he remains immobile in mother's lap, his mind moves far away to a never-never land of drifting clouds and heroic adventure.

Sign and symbol

The problematical relationship between environment and thought can be presented in another way. Environment may be viewed as systems of signs for the different animals who live in it. Each species responds to its own system or schema, although for related species there will be varying degrees of overlap. Signs are landmarks, more or less stable features in a scene, that enable animals to feel at home in their milieu and move through it effectively. Most of the signs are abstract to the animals who navigate by them: a tree stump is simply a signal in the visual field that says, 'turn left', and a boulder a signal for making a pause, and so on. Among the higher animals, some of the signs will have the power to evoke feeling or emotion. These signs have become less abstract by virtue of their affective content. One can imagine that an animal will have a special feeling for its nest, breeding ground, water pool, stable sources of food, and also for landmarks that arouse anxiety and fear. This world of the higher animals is, by and large, also that of human beings. We live and move in schemata of signs and affective signs. In such a world of patterned and routinized behaviour, the models of stimulus-and-response and of behaviourism can have a high degree of descriptive and predictive value. For human beings, however, the environment also contains symbols. And what are symbols? One way to define them is to say that they are constructs of *felt thought*. Consider, for example, the cross that appears on top of a church spire. Most of the time, it probably functions as a mere sign. When one sees the cross at a distance, it means that one is close to the shopping mart, or that one should turn right at the next street. Occasionally, it may function as an affective sign: that is, it will arouse a certain mood or feeling. Even more rarely, it will serve as a symbol: that is, it will create a mood or feeling that is a consequence of dramatic *ideas* – those of Christianity.

Under what conditions will the cross appear to an individual, not as a sign or even as an affective sign, but as a symbol? Will it be when the person approaches the church for morning service? Perhaps. But there can be no confident answer to this question. Behaviour itself, including facial expression, provides no definite clue. Of course, we can confidently say that the symbol response, in distinction to the other two types of response, is unlikely to occur when a person is on the run. There has to be a certain stillness for complex ideas and images to emerge. But this is a way of saying that, for human beings, precisely when environment does not elicit any kind of observable behaviour, something uniquely human and important is happening. Major decisions are usually made quietly and in solitude. The ideas that germinate then can have a large impact on future action. Thus, for example, under the influence of the cross functioning as symbol, a person may decide to lead a less worldly life. And we thus come to another major function of the symbol. Unlike the sign and the affective sign, which are necessary guides to routinized action, the symbol, itself a product of thought, makes further thought possible and this can lead to radical change. Thinking, in other words, makes for instability and directional change, whereas behaviourism and the stimulus–response model presuppose a more or less stable world – a closed system.[4]

Environment and communication

Environment and behaviour are closely correlated for a variety of reasons, which Roger Barker has specified. Einstein walks every day from his home to the Institute for Advanced Study and then back again. His observable behaviour is quite ordinary and can be inferred from the clues in his environment. What he thinks, which is at the centre of his life, is another matter. Yet what he thinks may, in the course of time, transform the world. Thinking is pursued in the privacy of the mind. However, it can also take a public form as conversation or dialogue. What is the relationship between environment and conversation – its themes as well as the manner in which it is conducted? We can confidently predict what people will do when they are in a restaurant: they will sit down at tables and eat. We can also predict what they will say to each other, though with far less assurance: they will talk most often about food. On the other hand, food may not be the topic at all. Despite the signs in the environment which all say 'eat', people may talk about the stock market or the latest advances in feminist critical theory.

If the topic of a conversation is hard to tell from environmental clues, it is even harder to tell the quality of the exchange. By quality, I mean both the courtesy and attentiveness that the participants extend to each other and the excellence of the ideas presented. Is there any relationship at all between environment and real communication, real exchange, and real enquiry? This is perhaps the key question for all ambitious planners, who would like to manipulate the physical setting in such a way that it promotes community. In a good community, people are able to talk with one another, courteously and effectively. In an ideal community, Glenn Tinder argues, people are engaged in free enquiry: they are searching for not only the common good understood as material welfare but also for 'the good,' for beauty and truth. As political beings and activists, people can create, with care and imagination, favourable physical and institutional con- ditions for communication and community. However, between the external arrangement, where action must terminate, and 'the communication that it is designed to produce there is a gap that can be filled only by spontaneity'. This gap requires what Tinder calls 'an interval of inaction' – or an interval of attentiveness.[5] During this interval, something *may* happen to the participants that enables them to rise to true enquiry and discourse.

Paradox of complete knowledge

Needless to say, we know practically nothing about what goes on in the mind during the span of reflective pause. We know very little, for that matter, about what 'attentiveness' means, for it is not only a psychological measure of focusing but also a moral stance: significantly, the literal meaning of respect is 'to look again'.[6] Finally, we know very little about the social psychology of the group when its members willingly submit to an interval of inaction. Suppose the time comes when we do have complete knowledge, when we are able to see in detail all the myriad links between environment and not only observable behaviour but sentiment and thought, then we shall be God. And we shall have God's dilemma or paradox – well known to theology – namely, how an omniscient and omnipotent Being can create beings with free will.

Notes

1 R. G. Barker, *Ecological Psychology* (Stanford University Press, Stanford, 1968), p. 29.
2 H. Arendt, 'Thinking', *New Yorker*, 28 November 1977, p. 126; see also George Kateb's review of Arendt's book *The Life of the Mind* in *American Scholar*, vol. 48, no. 1 (Winter 1978–9), p. 120.
3 P. H. Stephenson, 'On the possible significance of silence for the origin of speech', *Current Anthropology*, vol. 15, no. 3 (1974), pp. 324–5.
4 Yi-Fu Tuan, 'Sign and metaphor', *Annals of the Association of American Geographers*, vol. 68, no. 3 (1978), pp. 363–72.
5 G. Tinder, *Community: Reflections on a Tragic Ideal* (Louisiana State University Press, Baton Rouge, 1980), pp. 44–5.
6 Yi-Fu Tuan, 'Attention: moral–cognitive geography', *Journal of Geography*, vol. 86, no. 1 (1987), pp. 11–13.

Chapter five

Humankind–environment: musings on the role of the hyphen

Douglas Pocock

The introduction of the concept of behavioural environment is acknowledged among geographers as a benchmark in their study of people's interpretation of and response to the world around them. Subsequent work in perception, behavioural, and humanist studies are particularly indebted to the stimulus of Kirk's early paper.[1] Relationship lies at the heart of geography, whether our subject be considered an integrative discipline, a bridge-subject linking both physical and human or science and arts, or whether it be defined as regional description or in terms of humankind and environment. In Kirk's model the psychological ordering of the phenomenal or objective in the behavioural was a new articulation of this fundamental theme. The following brief foray is a personal reflection on the nature of this relationship or 'betweenness' signified by the hyphen in our study of humankind–environment relations.

Conventionally, the hyphen linking humankind and environment has represented relationship between two variables, one independent and the other dependent, one active and the other acted upon. The link is thus unidirectional. When the environment is inert and acted upon, then there are 'no necessities, but everywhere possibilities' and the human being 'as master of the possibilities, is the judge of their use'.[2] Powered by increasing technological competence, there is increasing impact on the globe for the benefit of humankind through continual discovery and conquest. The biblical injunction to subdue the earth may or may not be invoked. Any problem is deemed a challenge, not least any constraint imposed by the environment. Thus both Warntz[3] and Philbrick[4] confidently predicted overcoming what they independently described as 'tyranny' imposed in one instance by distance or space and in the other by place. The ultimate in our evolution is seen in what Cox termed 'technopolitan man',[5] a cultural being distinct not only from 'tribal' but also from 'urban man' enjoying freedom, anonymity, and maximum choice within the modern technopolis. Residences conceived as 'machines pour vivre' and lives played out in 'non-place urban realms'[6] are congruent elements of a picture where humankind is master of all that is surveyed.

The above interpretation reverses if our two variables exchange roles so that the hyphen signifies power flowing from the environment to determine the human response. Ecological, sociological, architectural determinism are variants of the same phenomenon as different disciplinary-defined environments posit humankind as a creature of circumstance, subject to forces beyond personal or collective control. Within geography perhaps the best-known statement is that of Semple; certainly there is no more dramatic opening than that to her *Influences of Geographic Environment*:

> Man is a product of the earth's surface. This means not merely that he is a child of the earth, dust of her dust; but that the earth has mothered him, fed him, set him tasks, directed his thoughts, confronted him with difficulties that have strengthened his body and sharpened his wits, given him problems of navigation or irrigation, and at the same time whispered hints for their solution.[7]

Although her case was too causally argued, with human passivity too easily assumed, and although environmental determinism fell from favour, nevertheless it has remained a concept more easily disapproved of than actually disproved. The very titles of modern journals, such as *Environment and Behavior* and *Architecture and Behavior*, suggest a continuing discussion. Certainly its complete dismissal would strike at the very core of geography. Interestingly, beyond our discipline a host of writers, both artists and scholars, have continued to acknowledge an active role of the physical environment over and beyond the obvious physiological adaptations to extremes of altitude or climate. Ecologist Dubos, for example, describes culture as not only a collective acceptance of certain conventions and traditions but also 'deep resonances with the qualities of nature from which they derive inspiration and sustenance'.[8] Novelist Durrell, similarly, considered spirit of place to be 'the important determinant of culture', adding with poetic conviction:

> Just as one particular vineyard will always give you a special wine with discernible characteristics, so a Spain, an Italy, a Greece will always give you the same type of culture – will express itself through the human being just as it does through its wild flowers.[9]

Poetic allegation is not dispassionate demonstration, but there can be little doubt that people differ by country or region, or that the adjectives Spanish, Italian, or Greek, for instance, are summary appellations denoting different life-styles, attitudes, and cultural productions. And it is not unreasonable to assume that peoples are shaped, not only by this human context, but also by their non-human environment as they daily and habitually respond to particular physical landscapes, buildings, weather. As individuals we are the sum of our experiences, all of which take place somewhere, and thus *who* we are is inseparable from *where* we are, where we were born, and where we have been. Where we come from forms one of the first items exchanged in any getting-to-know-you situation. In earlier times a person's name might well derive from the home place, the location of one's origin and life's experiences. Today, titles of non-hereditary earldoms echo the practice, but for Everyman, according to theologian Williams, place remains 'the medium by which its inhabitant feels and knows his own identity'.[10] Phenomenologist Norberg-Schultz considers that 'human identity presupposes the identity of place'.[11] The concepts of 'topoanalysis' (as an alternative to psychoanalysis) and autobiogeography follow naturally from acceptance of an active role of the physical environment.

Confirmation of the importance of the physical environment may be judged in people's reaction to separation or its destruction. Exile was feared by the ancients, homesickness may be experienced today. Immigrants and expatriates have characteristically reconstructed a reassuring homeland vernacular in their new countries. They are thus bolstered against what the French term '*dépaysement*' (literally, 'uncountried'). One modern example of a people *dépaysé* is the Bikini islanders, who were moved in 1946 to permit US atomic tests on their atoll home. After three relocations around the Marshall Islands, an official report partly attributed their continuing unsettled condition to their collective identity which persisted in regarding themselves as Bikinians and not Marshall Islanders.[12] In the Western world urban redevelopment, often justified as slum clearance, provides plenty of examples of what Fried called grief and fragmentation of spatial identity.[13] The challenge of geographical change, if experienced as 'future shock', can be sufficiently stressful as to be a factor predicting cardiovascular and depressive illness.[14] Little wonder, then, that the passage – or 'ravages' – of time evoke a seeking of stability or permanence in the physical. Individuals surround themselves with mementoes, nations actively preserve, restore, even rebuild a host of monuments – statues, ships, buildings, even whole settlements or town centres. 'Love places for the sake of good things' was Pope Gregory's advice to Augustine to confirm existing sites as points of worship during the missionary thrust from Rome.[15]

The role of the physical as a 'freezer' of events or marker of history is seen in the urge for pilgrimage in the human psyche. Recent publicized journeys which may be interpreted in this light are the anniversary visits in 1985 by ex-servicemen to the beaches of Normandy and to particular Pacific islands to renew the physical bond of critical experiences during the Second World War. (Many a small bag of beach sand was collected as a tangible reminder.) In the same year another party undertook a trip of some 8,000 miles to the Falkland Islands in order to see where their loved ones died and were buried in the more recent conflict.

The concept we call spirit of place also suggests something beyond human manipulation located in, or associated with, the physical: 'an independent expressive energy' is the definition by Walter.[16] Animists, pantheists, and oriental religions acknowledge or worship spirits in natural objects. In the modern, Western world, although spirits in the inanimate have been exorcised by Christianity and dismissed by science, there yet remains abundant evidence for the continued influence of this expressive force. At one level there are those who look to the stars to foretell their horoscope or who talk to plants (or tune into a plant's mood with a bio-activity translator). More general acknowledgement of something 'out there' is implicit every time we refer to nature having therapeutic qualities. Again, everyday speech suggests that we are all animists to a degree. We personify the environment or employ anthropomorphic descriptions. We admit to being moved or captivated by particular scenes, but what are such feelings if not a response, a *re*-action, to something already present?[17] Among poets, the 'genius of the place' of Pope or 'inscape' of Hopkins, among scholars, the 'god within' of Dubos or 'natural place' of Norberg-Schultz, among practitioners, perception of landscape potential by 'Capability' Brown or releasing of figures from marble blocks by Michelangelo – all bear witness to the recognition of an 'independent expressive force'. And all, interestingly, find an echo in the opening words of Semple of the earth having 'given problems ... and at the same time whispered hints for their solution'.

The above collage of examples suggests that humankind and environment are both active and acted upon. The hyphen, then, signifies a two-way link, a transactional relationship, a reciprocity. 'Mind takes form in the city, and in turn, urban forms condition mind', was Mumford's summary in *The Culture of Cities*.[18] More direct was Churchill's comment, 'We shape our buildings, and afterwards our buildings shape us.'[19] Behind it lay his belief that the nature of parliamentary debate, and hence British democracy itself, would be altered by anything other than an exact restoration of the Commons Chamber after Second World War damage. Reciprocity is at the core of the key concept of place, a focus which unites both the human and non-human. In the words of Ley, 'Place is a negotiated reality, a social construction by a purposeful set of actors. But the relationship is mutual, for places in turn develop and reinforce the identity of the social group that claims them.'[20] The mutual relationship described here is not a sequential or alternating, but a co-existing, force. In short, 'people are their place and a place is its people', and although the two may be separated in conceptual terms, 'in experience they are not easily differentiated'.[21]

An integration of two variables which are not easily differentiated, where each is both active and acted upon, might be represented by two hyphens, each denoting the link from the active to the passive. Such a doubling, of course, produces an equation signifying synthesis or union, which most obviously describes the monist world-view of oriental societies. The rigid dualism of Western societies, however, may be queried through an experiential perspective. (It might also be queried by materialist philosophies that consider humankind to be but part of nature.)

The most obvious or complete merging of person and environment in the Western conscience is that seen in transcendental or mystic experiences. What Seamon calls 'heightened contact'[22] and Ittelson 'environment as self'[23] is the state all may feel when 'one with' or 'wrapped up' in the world during moments of particular enjoyment or insight. In imaginative literature 'epiphanies' or 'permanent impressions' refer to fusing of person and place in the indelible recording of significant events. Our memory informs that it is not only major events that are embedded in place, for the 'where' is often an integral part of apparently insignificant happenings, as when having a 'clear picture' of where we were during a particular conversation or being able to solve the frustrating problem of knowing a face but being unable to 'place' it – until the appropriate context springs into mind.

Consistent with the consensus of theorists[24] who consider self a social construction and intersubjective is the concept of the social environment. Humankind constitutes environment in that we react to and with the presence and disposition of people. Accordingly, architectural or natural space *per se* does not trigger automatic or consistent responses. We experience, for example, that a room with one person changes character with the entry of a second, again with the addition of another two or three and once more with the presence of a crowd. 'It's the people that make the place', we may say. 'What is the city but the people?' was the comment of Sicinius in Shakespeare's *Coriolanus*. Little wonder, therefore, that disappointment is the frequent companion when revisiting early scenes long harboured as precious during the intervening years. In 'coming up for air', as Orwell found in Lower Binfield,[25] the traveller may well be moving in the uncomfortable and non-conformable coincidence of two worlds, with familiar surroundings strangely no longer offering the reassuring 'insideness' formerly known and assumed.

The related behavioural concept of social space holds that interpretation of the environment is culturally determined, but in that we personalize our surroundings, individually and collectively, then in experiential terms the environment is also imprinted with our identity so that it reflects us, it *is* us. The psychological and existential significance of the house as home is a widely acknowledged occurrence of this phenomenon.[26] Abundant examples exist in imaginative literature: Freud conceived the human personality to be grounded in the structure of a house; in a Jungian framework each floor represents a different level of human consciousness; Heidegger used linguistic evidence to show that dwelling and being had a common root.[27] Little wonder, therefore, that change or loss of dwelling can repercuss on our wellbeing. The effect of slum clearance has already been mentioned. The emotional effect of burglary has been likened to mugging or rape in that our private world has been violated. (The French word for burglary, '*viol*', literally means 'rape'.)[28] Even the misplacing of a treasured personal possession may be described in terms of losing part of one's self.

The integration of person and environment is brought into focus by workers who, treating reality as an experience rather than an object, question the reality of the body's epidermis as the boundary of self. The so-called 'phantom limb phenomenon', where a patient retains a sense of presence of an amputated arm or leg, is a macabre example where experience of self extends beyond the physical. 'Some thirty inches from my nose / The Frontier of my Person goes' is a couplet of Auden cited by Hall as an example of a proxemic pattern.[29] As a personal space surrounding each person it is a 'hidden dimension', and boundary of which we cross at our peril without personal invitation. Hall lists a series of proxemic patterns, which depend on activity and vary by culture, but perhaps the boundary of activities is better conceptualized as a gradient of involvement. Evernden argues in such manner, considering the individual as an intangible kind of force-field or field of care[30] and quoting Barrett who describes his Being as 'spread over a field or region which is the world of its care and concern'.[31] 'Field of care', interestingly, is Tuan's definition, not of person, but of place.[32] Noteworthy also is Heidegger's description of human being as 'Dasein', literally 'being-there' or 'that which regions'.[33] We thus turn full circle.[34]

Experience is unitary and care as a field signifies a bonding or betweenness. And it is care, or love, that 'makes the world go round'. Thus, when we say the 'world changes'; it is the betweenness, the relationship, that changes. The self in this sense is not a thing but the outworking of ways of relating. And since 'relation is reciprocity', an I–Thou relationship as described by Buber,[35] then it follows that the non-human also assumes meaning from context. In fact, it can be argued that nothing in the universe has any significance except in its connectedness.[36]

Relation or betweenness, then, is the characteristic of humankind – temporally, on the narrow isthmus of the present between the oceans of past and future; existentially, between a series of dialectical structures of our lived reciprocity; theologically, between this world and the next, with a 'Go-between God'[37] bringing disturbing comfort for believers. Geography, as the study of relations or betweenness, with a focus on contextual or hyphenated beings, is not the least among the disciplines worthy to engage in what is both a challenge for explication and a mystery for contemplation.

The above personal musings, as was acknowledged at the outset, owe their genesis to the early behavioural/phenomenal environment model of Kirk. Discussion here, however, has gone beyond a psychological ordering, or re-ordering, of the objective; beyond a cognitive approach in the study of environmental behaviour. Interestingly, Kirk himself, in a recent paper reviewing the evolution of his thought and placing his model within a broader philosophical framework, has emphasized a reciprocity between humankind and the world.[38] In his own argument and in the sources quoted, there is recognition that individuals both influence, and are influenced by, qualities of group structures and worlds in which they partake. The reflexive nature of a humanistic perspective, however, goes beyond cognitive processes and societal structures towards a more experiential focus. Experience – and resultant behaviour – is considered more than the sum of psychological, social, economic, and political forces. For much of the time individuals are immersed in the world with emotional and preconscious intelligence underpinning behaviour; certainly behaviour based on a continuously conscious process of decision-making is as unrealistic as the life so-led would be intolerable. From a humanistic viewpoint, a person-world or subject–object dichotomy or dualism is considered an intellectual concept out of touch with experiential or existential reality.[39] In contrast stands the phenomenologist's Being-in-the-World with taken-for-granted structures of his or her lifeworld. Relph has examined the degree of relatedness in terms of an insideness–outsideness continuum.[40] The view here has ranged more widely. Acceptance of its views and examples will hinge on the reader's belief system or world view, but, whether accepted or not, our relatedness remains.

Notes

1 W. Kirk, 'Problems of geography', *Geography*, vol. 48 (1963), pp. 357–71.
2 L. P. V. Febvre, *A Geographical Introduction to History* (Knopf, New York, 1925), p. 236.
3 W. Warntz, 'Global science and the tyranny of space', *Papers and Proceedings, Regional Science Association*, vol. 19 (1967), p. 7.
4 A. Philbrick, 'Perceptions and technologies as determinants of predictions about earth, 2050', in R. Abler, D. Janelle, A. Philbrick, and J. Sommer (eds), *Human Geography in a Shrinking World* (Duxbury Press, North Scituate, Mass., 1975), p. 33.
5 H. G. Cox, *The Secular City* (SCM Press, London, 1965).
6 M. Webber (ed.), *Explorations into Urban Structure* (University of Pennsylvania Press, Pittsburgh, 1964).
7 E. C. Semple, *Influences of Geographic Environment* (Constable, London, 1911), p. 1.
8 R. Dubos, *A God Within* (Charles Scribner's Sons, New York, 1972), p. 90.
9 L. Durrell, *Spirit of Place: Letters and Essays on Travel*, ed. A. G. Thomas (Faber & Faber, London, 1969), p. 156.
10 H. A. Williams, *True Resurrection* (Mitchell Beazley, London, 1972), p. 82.
11 C. Norberg-Schultz, 'Place', *Architectural Association Quarterly*, vol. 8 (1976), p. 8.
12 L. Mason, 'Kili community in transition', *South Pacific Quarterly Bulletin*, vol. 18 (1958), pp. 35–48.
13 M. Fried, 'Grieving for a lost home', in L. J. Duhl (ed.), *The Urban Condition* (Basic Books, New York, 1963).
14 A. Toffler, *Future Shock* (Pan Books, London, 1971).
15 Bede, *A History of the English Church and People* (Penguin, Harmondsworth, 1968), p. 73.
16 E. V. Walter, 'The places of experience', *Philosophical Forum*, vol. 12 (1980–1), p. 175.

17 J. Appleton, 'Environmental perception and the phoenix of animism', paper given to Geography and Literature Session, Institute of British Geographers Conference, Edinburgh, January 1983.
18 L. Mumford, *The Culture of Cities* (Secker & Warburg, London, 1940), p. 5.
19 R. K. Merton, 'The social psychology of housing', in W. Dennis (ed.), *Current Trends in Social Psychology* (University of Pittsburgh Press, Pittsburgh, 1948).
20 D. Ley, 'Behavioral geography and the philosophies of meaning', in K. R. Cox and R. G. Golledge (eds), *Behavioral Problems in Geography Revisited* (Methuen, New York and London, 1981), p. 219.
21 E. Relph, *Place and Placelessness* (Pion, London, 1976), p. 34.
22 D. Seamon, *A Geography of the Lifeworld* (Croom Helm, London; St Martin's, New York, 1979), pp. 111–13.
23 W. H. Ittelson, 'Environmental perception and urban experience', *Environment and Behavior*, vol. 10 (1978), pp. 202–3.
24 F. Johnson, 'The Western concept of self', in A. J. Marsella, G. DeVos, and F. L. K. Hsu (eds), *Culture and Self* (Tavistock, London, 1985), pp. 89–138.
25 G. Orwell, *Coming up for Air* (Secker & Warburg, London, 1948).
26 C. Cooper, 'The house as a symbol of self', in J. Lang, C. Burnette, W. Moleski, and D. Vachon (eds), *Designing for Human Behavior: Architecture and the Behavioral Sciences* (Dowden, Hutchinson, and Ross, Stroudsburg, Pa., 1974), pp. 130–46.
27 Norberg-Schultz, 'Place', pp. 8–9.
28 P. Korosec-Serfaty and D. Bolitt, 'Dwelling and the experience of burglary', *Journal of Environmental Psychology*, vol. 6 (1986), p. 338.
29 E. T. Hall, *The Hidden Dimension* (Doubleday, New York, 1966).
30 N. Evernden, *The Natural Alien: Humankind and Environment* (University of Toronto Press, Toronto, 1985).
31 W. Barrett, *Irrational Man* (Anchor Books, Garden City, NY, 1962), p. 217.
32 Yi-Fu Tuan, 'Space and place: humanistic perspective', *Progress in Geography*, vol. 6 (1974), pp. 211–52.
33 M. Heidegger, *Being and Time* (Harper & Row, New York, 1962).
34 G. Olsson, *Birds in Egg* (Pion, London, 1975).
35 M. Buber, *I and Thou* (Charles Scribner's Sons, New York, 1970).
36 D. Hayward, 'Man blinded by choice', *The Times*, 22 November 1986.
37 J. V. Taylor, *The Go-between God* (SCM Press, London, 1972).
38 W. Kirk, 'The road from Mandalay: towards a geographical philosophy', *Transactions, Institute of British Geographers*, New Series, vol. 3 (1978), pp. 381–94.
39 D. Seamon, 'Philosophical directions in behavioral geography, with an emphasis on the phenomenological contribution', in T. F. Saarinen, D. Seamon, and J. L. Sell, 'Environment perception and behavior: an inventory and prospect', Research Paper no. 209 (1984), Department of Geography, University of Chicago, pp. 167–78.
40 Relph, *Place and Placelessness*.

Part three

Application

Chapter six

People, prejudice, and place

Wreford Watson

Many geographers have found their thinking echoed in William Kirk's work on the role of culture in the development of the landscape. His 'eye of culture' sees what it chooses in the land and turns the land into what it wants. For years I, too, have tried to show that the images conjured up by the land create an *image geography* that then determines how the land will be developed.[1] In pursuing these ideas I owe a great deal to William Kirk and am happy to join in this collection of essays on the theme of the behavioural environment. My wife, too, honoured his ideas in her concern with teaching the cultural background to human perception of the earth. In her paper on 'The teaching of value geography' she wrote, 'Geographers . . . are involved in people who have unique views and feelings. Based on these views and feelings, people develop values, and based on these values they make choices. These choices affect the earth and change the landscape.'[2]

The fundamental role of the behavioural environment is clearly exposed in the attitudes of people towards place. In this chapter, therefore, I want to reflect on the prejudices people have for the kind of place they would like to live in because these are amongst the most powerful forces shaping the rise and growth of settlements. Indeed, whole hierarchies of towns have grown up on the strength of such prejudice. Furthermore, there has been a widespread restructuring of hierarchies as people's tastes and prejudices have changed.

First, I should like to deal with the *general* changes in urban hierarchies, from medieval days, through Georgian and Victorian to modern times, as instanced by the four major Scottish cities – Edinburgh, Glasgow, Aberdeen, and Dundee. Second, this story of change will be looked at more *specifically* in terms of Edinburgh. Finally, brief conclusions are put forward about hierarchical restructuring which, though derived from Britain, might apply elsewhere.

General changes in the urban hierarchy in Western Europe started from 'the model of the enclosed city which has now become so much of a dinosaur that it is in danger of killing itself'.[3] Cities closed around what they put their faith in, namely the wall-building institutions of Crown and Church. The medieval prejudice for king and bishop as the pillars of society prevailed, often with a strong personal attachment to the Crown. The Crown was the cornerstone. From it sprang the prominence of many a royal burgh that has long ago lost its importance. In Scotland, medieval loyalties, where men hardly attempted to think except by 'permission from a monk or an officer',[4] were kept alive till as late as 1745! The passionate prejudice of the Roman Catholic Highlanders for the Jacobite cause and for the return to the throne not only of Bonnie Prince Charlie but of the Divine Right of Kings maintained the medieval mentality well after it had gone in England – with Charles I's head!

Nevertheless, this mystical attachment to the old ideals *did* decline, and many royal burghs and holy sees fell by the wayside, with the rise of a new urban hierarchy. Here, the Merchant Hall challenged castle and cathedral as the new node of existence. Then places like Glasgow, which had not been one of the original four or five royal burghs of Scotland, surpassed the lot, even including Edinburgh – the leading home of royalty and capital of the kingdom. Fortunately for Edinburgh it was better placed for the hierarchy of business than most other royal burghs (for example, Dunfermline), which, after their days of note in the medieval hierarchy, sank to comparative insignificance, and Edinburgh was able to continue as an active business centre.

This resulted in the emergence of the *central business district*, a feature that soon dominated the internal structure of the Georgian, and later the Victorian, city. In Edinburgh it was centred on Princes Street and caused the complete split between the medieval town and the Georgian New Town. Glasgow, likewise, saw a shift from cathedral, college, and the old city centre known as Glasgow Cross to New Town Glasgow, around George Square. In Aberdeen the breakaway became a tearaway and what was left of the medieval town around its cathedral and college is known to this day as Old Aberdeen: a new town axis grew up as High Street was abandoned for Union Street, the central business district of new Aberdeen. Only in Dundee did new and old have to get on together because local geography gripped the town site as in a vice between Dundee Law, a high volcanic peak to the north, and the Tay estuary to the south, yet to make room for a new broad avenue, like Commercial Street, the old city walls and the castle with them had to be knocked down.

Internal structures changed even more in Victorian times as the prejudice *for* business led to the virtual eclipse of the Divine Right of Kings (and of castle and palace as central urban institutions), in favour of the right of the market and of finance, wholesale and retail business, and industry, as the makers and shapers of the town. With the better opportunities Glasgow had for trade and industry, facing as it did the North American continent and the rise of Canada and the USA, Glasgow far outstripped its Scottish rivals and became the Second City of the United Kingdom – or, for that matter, of the British Empire. The castles went in Glasgow, Dundee, and Aberdeen, and even the famous Edinburgh castle was downgraded, in spite of its physical prominence, to give place to the railway station and the station hotel as the main centre of the town.

Indeed, the prejudice in favour of vehicular mobility, of train, bus, and finally automobile, led to the replacement of many features associated with pedestrian town circulation. The modern-style business city began to regard it as sheer 'pedestrianism' to keep walls and gates intact, or to allow cemeteries a central location. The holy ground around St Giles Cathedral in Edinburgh was de-hallowed and built over by secular institutions: even the bones of John Knox were uplifted, and nobody knows to this day where they lie! Furthermore, Edinburgh's famous Netherbow Port, the great gateway between the upper and lower towns of medieval times, was torn down in 1764, just when the Georgian New Town was about to rise.

However, the rule of business, strong though it was, came to be challenged in contemporary times, by the claims of the community, as the *social* use of land challenged *economic* land values. As a university student I was taught by the political economist, Sir Alexander Gray, that nothing could be socially stable that was not economically sound. This was asserted as the *sine qua non* of human progress. Hence, walls and gates had to disappear in favour of trade and traffic. Today we have reversed all that – for nothing is economically permissible which is not socially desirable. This is the most powerful city-shaping prejudice of our time.

The whole value-system has been called into question: land values are no longer determined by strict business profitability (by what used to be called the economic rent), but give way to social preferability – a matter very largely based on community prejudice. That prejudice favours a more egalitarian society which has, for instance, seen low-income-level people move out into their council-housing flats and enjoy the suburbs of light and air, or bird-song and greenery along with higher-income families. Here is a tremendous revolution in the landscape, which has grown out of a revolution of the mind. And those towns which are perceived as socially the most desirable to live in – giving light, air, bird-song, and greenery to most of their people – have become the 'desire-line geography' of today. Families choose their geographical lines of movement by socially desired goals. A new urban hierarchy is being born.

Thirty years ago Osborne and Jones studied the desire-line geography of would-be teachers who had just graduated from college.[5] They found that for most of them the last place in which they desired to make their life was Glasgow, since its image was of an industrial town, grimed with the 'ugly face of capitalism'. One should hasten to add that the immense efforts of Glasgow to clear its slums, beautify the central city, and develop low-cost housing at its suburbs, much of which is on hills that give to even the poorest of its citizens magnificent views down the Clyde or up to the Scottish Highlands, have changed the situation. One is not sure, however, from the extreme shortage of teachers in Glasgow, that the old image does not remain!

More research on the changes in urban hierarchies due to changes in people's minds must be done. It was the stimulus of Michael Ray, of Carleton University, that led to this particular discussion of the Scottish situation. While he was a visitor at the Centre of Canadian Studies at Edinburgh University he kept asking what was it that produced the dramatic (and also traumatic) change from Old Edinburgh to Edinburgh New Town. What made towns constitute themselves into new hierarchical orders? To what extent did this alter not only their external fortunes, in relation to scales of urban size and influence, but their internal features, as they adopted the key institutions of the changing hierarchy?

To say that this is a matter of changing economic functions is not enough. To put it down to new social modes is not sufficient. Who thought up a new economic order while an existing one was still in power? By what sort of talk were people's set prejudices for one social system so altered that they became biased towards a new one? People have to be *moved* to make a change. What were the movements that reshaped their minds? 'People' are, of course, made up of persons: what persons became the midwives of change? Marwyn Samuels has argued strongly for the biography of landscapes, meaning the biography of those who actually pre-cipitated landscape change.[6]

For the fact is, though movements make people, people create movements: both processes are involved. Geographers have per-suaded themselves – perhaps for too long – that people in the mass are our concern; we tend to eschew the individual. However, people in the mass so often follow a person, like the Highlanders and Bonnie Prince Charlie, or the Lowland business community and Adam Smith. And where people-in-the-mass change their thinking, more especially where they do so almost overnight, one can scent out a person. Up to 1745 Scotland was afraid to make the changes that had begun to be accepted in England during Queen Elizabeth's day: yet by 1765 the plan for a new Edinburgh had been adopted. In those 20 years a tremendous sense of liberation arose: the old régime was gone with Bonnie Prince Charlie (the symbol of feudalism), a new one was called for by Adam Smith (the father of capitalism). In fact, a change nothing short of a catastrophe had occurred for the old system, allowing a new landscape to burst upon the scene.

Changes in urban hierarchies often occur like that – suddenly, through a great leap forward, as a unique act inspired by a unique leader or group of leaders. Both Glasgow and Edinburgh shared in the men that gave point to those movements producing new towns from old, particularly in Adam Smith who called for the freedom needed to make the change, and suggested the felicity with which the change was made. In Edinburgh, support was also lent to the new ideas by Hume and Ferguson, and by the dynamic Lord Provost of the day, Drummond.

Turning from generalities, it will be profitable to look more closely at Edinburgh; for here, the leap from the old city to New Town was particularly dramatic. Medieval Edinburgh had grown up as a system of cells centred on castle, cathedral, and palace with a system *within each cell* of markets, work-places, schools, chapels, and people's homes. Each cell was virtually a separate community, although each was tied to the other by one high street (subsequently known as the Royal Mile), the union of Church with Crown, and the need for a common defence.

Since each major institution was different in function, its cell-like structure differed in form. Take, for example, the cell of the castle. Its houses consisted of billets or married quarters for the garrison; saloons were frequent – celebrated by the soldiery; a Grass Market and the King's Stables road were unique features – for the horses of cavalry and horse-drawn wagons; as were the alleys for armourers and ironsmiths. The cathedral was linked with the houses of the clerics, with the 'luckenbooths' or stalls for jewellers, robe makers, booksellers, and 'lawn' or linen weavers; scribes or 'writers to the signet' – solicitors and advocates, and the law courts were also nearby; so too libraries, schools, and almshouses. The palace, in its turn, created a cell of courtiers' homes – the town-houses of great landed families with seats in the far parts of Scotland; also with breweries and wineries to keep the entertainment flowing; fashion shops (where for example the first wig was sold); inns for travellers from England or abroad; and, not least, the royal bath-house!

In this kind of institutionally dominated town, there was no central business district. There was no overall segregation of housing by land values. Separate markets served each cell, and there was a great social mix of aristocrats, clerics, merchants, craftsmen, and the *hoi polloi* in each neighbourhood. This led to quite idiosyncratic arrangements of space: the castle mart was distinctly separate from, though immediately below, the castle; the cathedral market flowed all around it and, in the form of the notorious 'luckenbooths', actually impinged upon it; the palace market was its front yard. Each market had quite irregular entries and gates; each was made up of temporary booths or permanent buildings of quite irregular width, height, depth, and appearance. The freedom which the medieval merchant wanted was to secure the special favour of the main institution he served (if possible by royal or episcopal appointment) and to get a spot as close to this as he could. Hence he thrust his building forward into the street, or built galleries out over the street, or got the street to make a turn or a recess in his favour, or lured people off the street down narrow alleys to back courtyards, or laid his goods right out in the streets. Even 'vegetables were laid down in the street and on the footpaths right beside gutters running near them in the most offensive way'.[7]

There seemed to be little that was felicitous; and yet, in the eighteenth and nineteenth centuries when much was done to straighten out and clean up the Old Town, many sentimental citizens opposed change simply through their prejudice for the idiosyncratic. People had liked the uneven; they had had a taste for the irregular. This was supported by the favouritism which kings showed to their courtiers and the privileges these passed down to their hangers-on; or with the protection the Church could offer its followers – especially those whose trade filled its coffers. Once, a king's man blocked an old right of way with an extension to his house. The town council objected, but the king set the council aside.[8] Both king and Church wanted to collect as high a revenue as possible off the land they owned, and therefore allowed it to be built on by individuals to the best of *their* advantage. This led to terrible overcrowding in which buildings along the High Street were raised taller and taller – 10-storey tenements were not uncommon, and the back gardens, or burgage lots of all the town's burghers, became filled in with building until the alley-ways wound through one back courtyard after another, like a rabbit warren.

There was no space left for children to play in, except the narrow alleys and back yards: green and open space had been virtually abolished. The space left to women, even for hanging out clothes to dry, was next to nothing: clothes were hung out on poles extended from windows high above a street or yard. Space for burning rubbish or disposing of garbage was not thought of. At dusk, women withdrew their washing from the windows and, leaning out, threw their garbage into the street, crying loudly 'Gardez lou' – take guard, below.

Adam Smith and David Hume both had experience of this, since both lived in the Old Town, although Hume moved to the New Town as soon as he could, and Adam Smith would have followed him, but for his untimely death. Undoubtedly, their medieval surrounds must have influenced them. Adam Smith spoke up sharply against Crown and Church, and Hume against the mystical thinking at the basis of their power. This was typical of Georgian times; and it resulted in a secular city planned for symmetry and order.

Georgian Edinburgh was an outcome of the Age of Reason. Although it thought itself unbiassed – it was after all, in Voltaire's words, out to 'find a truth which belongs to all time and all men', it had an almost pathetic prejudice for mathematics and science. People talked of a science of economics and a science of politics. Adam Smith was praised for having worked out *principles* for political economy. These same principles were seized on with enthusiasm by 'some very eminent merchants' who, according to Dugald Stewart, the first biographer of Adam Smith, 'adopted his system with eagerness ... and diffused a knowledge of its fundamentals throughout the Kingdom'.[9]

One of Adam Smith's main principles was to make a more *economic* use of land. He attacked the medieval system of treating land as a source of *revenue* instead of as a means of *profit*. The Crown and the Church were in effect sucking the blood out of the land, and not making land the bloodstream of new wealth, through manufacturing and business. 'In those towns', Adam Smith wrote, 'which are principally supported by the constant or occasional residence of the Court [as was the case with Edinburgh], and in which the inferior ranks of people are chiefly maintained by the spending of revenue, they are in general idle, dissolute and poor'. By contrast, 'In mercantile and manufacturing towns, where ... people are chiefly maintained by the employment of capital, they are in general ... thriving, as in ... Dutch towns.'[10]

Free enterprise was the answer, where 'the desire of bettering our condition, a desire which ... comes with us from the womb, and never leaves us till we go into the grave ... puts into motion the greater part of the useful labour of every society'.[11] Proposing that land should be valued for the wealth it could *produce*, and not *consume*, Adam Smith formulated a theory of land value based on the competitive bid for its economic use. This would make the point of maximum accessibility in any region the bid peak, and so concentrate business in the city centre. Here was the reason for a central business district: hence Princes Street, in Georgian Edinburgh, concentrated upon itself, and has retained to this day, the trading functions of the city, replacing the many and scattered markets of medieval Edinburgh. The signs of a new urban hierarchy had been born, based on 'economic man'. The days of 'mystical man' wrapped up in Crown and Church were over.

One might have thought that the immense sense of personal freedom unleashed by the capitalist system would have led to an even more irregular and individualistic city than before; yet this did not happen: another prejudice was at work. Edinburgh New Town saw one of the best-planned, symmetrical, and highly disciplined styles of city development that could have been imagined. This was because it grew up in the age of enlightenment, when freedom went with felicity.

It should be remembered that Adam Smith was a Professor of Moral Philosophy and of Aesthetics before he got the Chair of Political Economy. Smith's ethic rested on 'moral approbation or disapprobation of the actions of others brought home and applied to ourselves'.[12] No one could earn the approbation of his or her neighbour in pushing his or her house into the street in front of that neighbour. Hence, the freedom to take up lots in the New Town ended up in having all houses in the same line and height, and thus avoiding the disapprobation of any one burgher getting an advantage over any other.

In his *Aesthetics* Adam Smith talked of a systematical beauty as most to be admired. He laid stress on regularity, balance, conformity, and uniformity – all very much the distinguishing features of New Town Edinburgh. When he was a student at Glasgow he was especially good at geometry, a fact which Dugald Stewart claimed 'may be marked in all his reasonings'.[13] Smith said bluntly that the variety of shape and form in Gothic architecture 'is not agreeable'.[14] What he considered as agreeable was 'the exact resemblance of the corresponding parts of the same object ... as in the opposite wings of the same building', as for example in the Georgian-styled Registry House.[15]

Adam Smith was given strong support by David Hume who spoke of geometry as beauty, and also of the beauty of what was accurate. He too abhorred the fuzzy, the irregular, the wayward, the mystical, and the idiosyncratic. He wanted to give system and harmony to thought. This strong prejudice among the leaders of the Enlightenment led to a very geometric layout of Edinburgh New Town in rectangles and squares with broad boulevards and ending in fine views. There was no obstructive building widening or narrowing the streets, no squares with unseen and unanticipated openings, no houses jutting into the street or leaning over it from above, no sudden changes in colour or texture of building, no mystical symbols projecting like gargoyles from building façades, no windows of uneven height placed at uneven levels – nothing in fact that typified the Old Town: only the new rationality expressed in the new uniformity.

The Old and the New clashed in people's minds, and competed for their loyalties. As the Old Town became denigrated, denuded, and even abandoned, those who loved old causes sprang to its defence: as the New Town grew more progressive and efficient it developed *its* champions, in turn. Half way between the two movements, James Wilson, the town poet, cried in the 1780s 'for a spirit of reverence for the city's ancient edifices', and inveighed against

> the lack of taste in most of the new streets going up. In too many cases the quaint and picturesque Scottish domestic style, with its ruggedness and variety (seen in irregular gables and varied walls, and windows of different size and height), is being displaced to make way for straight lines and a dull monotony – where the same size of window is made to do duty throughout the whole length of a street.[16]

The predilection for straight lines and the same size of window was precisely what inspired Lord Provost Drummond to press ahead for the New Town, the style of which was to be so uniform as to 'prevent the inconveniences and disadvantages which arise from carrying on building without regard for any order and regularity'.[17] Obviously, here was a case where prejudice made place!

Another prejudice of the New Town was to keep it for people of means. According to Lord Provost Drummond's *Proposals*, setting out the ideals for the New Town, it was to become at once 'the centre of trade and commerce, of learning and the arts, of politeness and of refinement of every kind'. It was seen as the choice of 'people of fortune and of certain rank'.[18] It gave a chance for people of distinction to separate themselves out from the social mish-mash that characterized the Old Town. The prejudice to be apart created a place apart. Social distinction demanded geographical separation. The geography of élitism became a major factor in the new urban hierarchy. This was of course the hierarchy of the commercial city. As the New Town Proposals argued, the time had come when wealth no longer went with the solitary grandeur of a country seat suited to the stateliness and pride of gentlemen acting as petty sovereigns, but wealth was only to be obtained by trade and commerce, and these were only carried on to advantage in populous cities. This echoed Adam Smith who argued that it was only by 'the improvements of arts, manufacture and commerce' that 'the power of the great barons and ... the whole temporal power of the clergy' could be displaced.[19]

Thoughts like these helped to bring about the catastrophic change in the urban hierarchy seen in the swift collapse of baron and bishop as the key to town status, and their all but revolutionary replacement by merchant hall and bank. Merchant and financier were about to take over. Adam Smith forecast the rising pre-eminence of the commercial town: this was where anybody would flock to, who wanted to be *somebody*! 'Private people who want to make a fortune', he wrote, 'never think of retiring to the remote and poor provinces of the country, but resort to ... some of the great commercial towns.'[20] Drummond's *Proposals* for Edinburgh New Town underscore this. 'Wealth is only to be obtained ... in populous cities', he proclaimed. 'There also we find the chief objects of pleasure and ambition, and there consequently all those will flock whose circumstances can afford it.'[21] The flocking of the would-be rich into Georgian Edinburgh and then Georgian Glasgow certainly helped these places to a higher order of rank among Scottish cities, and put them well above those still dominated by baron or bishop. Indeed, not a few barons accommodated themselves to the change, and became businessmen!

Elitism was reflected in New Town geography by much more *personal space* – especially for women and children. Female space showed itself in more rooms within each house for dressing, bathing, sewing, and, above all, social entertainment. It also embraced the garden which the wife tended to oversee; and it made itself felt in the very fabric of the New Town, through shops specially designed for women – female-fashion shops of all kinds, and also cafés of taste and elegance where women could meet for coffee and cards. Child space was also enormously increased. Both Princes Street and Queen Street and all the new-town squares had private gardens facing them where nannies could take out the children and look after their play.

Thus, by comparing old medieval Edinburgh with the new Georgian town we see how the prejudices of people were played on, either to keep places as they had grown up in the past, or to help create the places of the future. Writers, officials, developers, romanticists, rationalists, those who identified themselves with the mystical values of royalty and the Church, or those who preferred the new economic values based on private enterprise (and all the light and space that went with them), all sought to change the minds of their fellows in favour of one hierarchical urban system or another.

Jay Appleton has the words for it: the Old Town was a refuge – a hide-out – for old causes; the New Town was a prospect – a look-out – for new enterprises.[22] This change went with a major hierarchical restructuring. The old hierarchy was one of towns that looked inward; they were highly centripetal, and they were structured in small cells centred on essentially social institutions. People in them were hidden by walls, protected by gates, cut off by alleys; they faced into courtyards; and they were led away by steps, and blocked by irregular streets and outward-projecting houses from any view of the outer world. Occasionally they got a peep-hole view of further dimensions, usually from the end of a 'close' down to another close; but they lived in a town of terminal vistas, and restricted circulation. One had to walk – or be carried in a sedan chair – about town, to appreciate the little deviations, recesses, byways, up or down flights of narrow stairs, overhead galleries, and gable-end roofs – almost all of them reflecting the prejudices of one landholder or another.

The new hierarchy was set out on the geometry of a grid system, marked by straight streets with no blockages to the view: it was regular and open. It had no unexpected turns, cut-offs, recesses, alley-ways, or hidden courts. If the great main boulevards were intercepted, they were intercepted in a regular fashion by side streets which were themselves broad and open. Whereas the Old Town had shown little sky, here the skies came right into, and were part of, the New Town. It was, in fact, a town of light and air, going out on all sides to magnificent views. It carried the town into the country: it brought the region into the city. There was nothing that was idiosyncratic; symmetry and uniformity appeared in the alignments, fenestration, and roof levels of all houses, in their style of architecture, and even of the material with which they were built. A common felicity was the order of the day.

To get a sense of this – of the balance struck by St Andrew's Square facing St George's Square a mile distant from each other along the central axis of Edinburgh New Town (George Street) or

103

the geometric symmetry of Princes Street and Queen Street running parallel to each other, each looking out on a parallel strip of gardens – one had to ride or drive round by horse and carriage. The New Town was made for mobility. The pedestrianism of the Old Town gave a blinkered view of life which went, of course, with a hierarchy of inward-looking institutions like palace and parish well content with their world. The equestrianism, or rather voiturierism, of the New Town, exposed people to space, it led them on and out, it opened up the world to them, a world which they would explore by commerce, and exploit by industry, and thus work into a new urban hierarchy.

The Victorian city had both inward- and outward-looking structures. It was made up of industrial cells each looking into the railway yard and the factory; but it was also developed looking outward along great axial routes to the suburb and the commercial hinterland beyond. Starting off as it did from the Georgian base, it continued to underwrite élitism, yet at the same time it produced Britain's largest working class. In fact, it accepted a class structured society, and it entrenched class division in the new urban hierarchy. Its élitism, however, became eclecticism; it borrowed whatever ideas – mystical, rationalistic, humanistic – that were useful to it. Swiftly the symmetry and uniformity of Georgian Edinburgh were changed. French baronial towers ornamented buildings with a Corinthian portico; Gothic windows relieved Palladian walls; Rococo iron grill-work decorated neo-classic façades. It seemed as if any style would do as long as it was done in a striking way: above all, as long as it pleased private taste.

Privatism dominated. More and more, Victorianism interpreted the perfect liberty Adam Smith had proposed as private liking. Although it continued the Georgian tradition of row-houses, it embellished this with rows of flats having Greek architraves over window and door, Romanesque balustrades, and Château-Loire turrets to give an air of opulence or grandiloquence to even middle-class terraces. Privatism also went with the growth of the private villa in the new urban hierarchy, with large private gardens surrounded by high walls topped with iron spikes or broken glass to ensure privacy, and private drives to the front porch with service entries to the back quarters – confirming an apartness that tore the community apart.

At the other end of society were workers' districts around the factories built about railway freight-yards. Here were juxtaposed, in the most crowded way feasible, warehouses, factories, coal dumps, railway sidings, spur lines to individual factories, draymen's yards and supply depots, workers' tenements, and local primary schools – to which corner shops and saloons, back-alley missions and labour-temples, and working-men's clubs were somehow added. It did not take long for these to deteriorate into slums, and make themselves prominent in the city police records.

Since the *main* railway lines tried to get right into the heart of the city and ensure that the railway hotels dominated the central business district, while the *branch* lines bracketed the rest of the central area with freight yards that brought industry as close to commerce as possible, an immense squeeze, a veritable bear's hug, was put around the heart of the city. In some cases parts of the medieval fabric and even Georgian terraces were cut through to make way for the railway and railside industrial areas.

The railway was the basis for industrial development. It was both bonanza and catastrophe. The railway invaded the fringes of the central area with factories and tenements, making a sharp break between the city and its suburbs; meanwhile the inner edges of the city centre saw the outward invasion of commerce. Invading forces shattered the living quarters in the downtown and mid-town areas, making them a zone in constant transition, and led to a flight to the suburbs as fast and as far as the suburban passenger line, the horse-drawn bus, cable cars, electric trams, and the automobile could carry the well-to-do.

Privatism led, then, to a hierarchy of towns built up and directed by the middle class where an outer-city affluence strangled an alienated inner city. Describing Victorian Edinburgh inside its ring of industrial sidings, Elspeth Graham claims, 'Here, in 1850 were some of the most unsanitary and densely populated streets in Europe.'[23] Such conditions brought their own reaction. People became increasingly prejudiced against privatism, and demanded a community say in affairs. Privatism, it is true, did try to offset the ugly with the fair face of capitalism. Private money built Edinburgh's main Concert Hall, the University's new Convocation Hall, and the Church of Scotland's Assembly Hall; it erected the Edinburgh Academy and Fettes College – two of Edinburgh's most famous schools, and made the ground of a great private house the basis of the Royal Botanical Gardens. In these and other ways it beautified Edinburgh; but mainly middle-class Edinburgh. Little was done about the slums.

Today's town grew out of the longing for a more socialized hierarchy of settlements. The strong and emotional prejudice against slums and Britain's 'dark satanic mills' blighting the homes or lives of millions changed the feeling about places. Places should be where all people could live with dignity and with justice. And so movements to reshape the city began to arise, and protesting voices were soon heard. One of the strongest was that of Robert Owen, whose prejudice for a socialized resolution led to the rise, in Scottish urban geography, of his planned mill-town, New Lanark. Here, protest had taken form in stone and lime. It was as real as that. People's aspirations for a new way of life had led to the creation of a new place. *Ideas* shaped the landscape; a *person* fathered a new hierarchy. However, it was a hierarchy at many years' remove. Owen's experiment of a socially planned town, on community lines, was not taken up. Nevertheless, the seed was sown.

Later in the century Patrick Geddes took up the planned-city idea. A personal friend of the great French free-thinker and geographer, Elisée Reclus, and of the expatriate Russian reformer, Prince Kropotkin, Geddes grew up on the verge of social revolution. In the early twentieth century, he became the first Professor of Sociology in Scotland, at Dundee College, a post he used in order to make social surveys and to teach the need for social reform. His mentor, Reclus, believed that humankind was the conscience of the earth and that it was up to civilized people to 'give the landscape charm, grace, and harmony.' People should assume the responsibility of beautifying their environs, and especially of returning the beauty which vicious exploitation had caused to disappear.[24] These ideas profoundly influenced Geddes, who tried to whip up the conscience of his country over the vicious exploitation which had marred its cities.

Like Owen, Geddes was not content with preaching. He formed the first working-man's housing co-operative in Edinburgh to restore badly deteriorated tenements in the upper High Street. Here, the prejudice for socialized space led to a new impact on urban geography. He called this 'conservative surgery'.[25] More, Geddes gathered money from his friends to set up the first Student's Co-operative Hall of Residence, planned and run by students for students. Its splendidly romantic building soars from the Castle Esplanade in Edinburgh Old Town to give a wonderful view over Princes Street Gardens, and New Town Edinburgh. Geddes fathered other co-operative projects, but he was increasingly aware that this was just a patchwork approach. What was needed was complete urban renewal: more than that, a renewed society.

This society must be based on communal needs, and on mutual aid. Kropotkin had urged that mutual aid and not the survival of the fittest was the way to further evolution. Individuals must limit their freedom to go after their own ends, so that they might have a greater chance of survival in the *freedom of the group* to achieve *its* end. There is a freedom through limitation as well as through assertion.[26]

Meanwhile, of course, in England that great architect of the Garden City, Ebenezer Howard, had been at work, urging people to challenge the private city by building new towns, planned on community lines, where living and working could both be beautiful. His famous plea was for 'the order of justice, unity and friendliness'. He once wrote:

I went into the dark crowded streets of London and as I saw the wretched dwellings in which the majority of the people lived and observed on every hand the manifestations of a self-seeking society, there came to me an overwhelming sense of the entire unsuitablity of all I saw for the working life of the new order – the order of justice, unity and friendliness.[27]

All these things added up to an irresistible urge for change, but it was not until the catastrophe of the First World War struck, and, shattered by that experience, people *so* wanted a better world, that the first Act was passed to plan British towns for a better environment. With this first Planning Act in 1917, a new hierarchy was on its way: the hierarchy of socialized towns where it was hoped all people could live better lives.

By coincidence, Britain found itself in a new Georgian era. This era was reminiscent of the first one in its insistence on geometry, on balance, symmetry, and regularity, although it based its design on the curve rather than the straight line, the circle rather than the square – that is, on streets that bring you back upon yourselves, and make a charmed circle of good neighbours. However, the second Georgian era was quite different from the first in that it was not for the élite but for everybody. The aim was to give the poorest and the lowliest the same access to suburban living among trees and hedges, with clean air, bird-song and light and views of surrounding hills that the richest in the land might obtain. Here was the hierarchy of social ends, of towns which would rank themselves accordingly as they furthered the social weal.

Contemporary Britain is seeking a better life by creating a better environment. Tree-lined boulevards, grass verges, service roads bleeding off the main thoroughfares, local shopping centres, a variety of housing from high-rise apartments for young couples to low-level houses for old folks, schools with large playing fields, youth centres, health clinics, and in many instances a small industrial estate, with landscaped factories, environ the people in suburbs of public amenity and public enterprise. Cities are buying up private land to hold as a public land-bank for industrial estates, community housing, and recreational projects. Their plans are put to public hearings, and tenant associations have their say in the public running of the eventual developments. In a city like Dundee, once ridden by private landlordism, 76 per cent of all the housing stock consists of low-cost public housing. In Glasgow about two-thirds of all housing is publicly owned, and in Edinburgh 54 per cent. These figures clearly show how far the contemporary urban scene is socialized: private ownership accounts for less than half the lands and building of the major cities of Scotland. The bias is against private property and in favour of public ownership.

Most of this development has been at the peripheries and has thus brought middle-class suburbs next to council housing estates, in a cheek-by-jowl relationship that is supposed to intermingle the classes, and get rid of class divisions. Were this to happen it would be a major change in British society, which witnessed such a class-segregated geography in Victorian times.

Another means of recreating the social mix is an urban renewal programme that will bring the middle class back into the inner city. This is a main feature of the master plan drawn up for Edinburgh by Abercrombie, one of Scotland's greatest planners. By rebuilding the slums *in situ*, and providing for apartments of different sizes and styles at different levels of rent, both highly subsidized homes for the poor and homes for better-off people at economic rents, people of different occupations and outlooks will be drawn together. This has undoubtedly helped to write off the conditions that had grown up under the hierarchy of privatism, based on economic land values. The emotionally felt prejudice people now have for social values (i.e. the lower class slap up against the upper) will revalue place and reshape the geography of the future. This is something that people strongly believe. That belief is translated into the reality of a changed scene. In this scene, things that are still cherished and relevant from the past are caught up into things that are dreamed about and worked for in the future.

It should be obvious from all this that geographers should lay more stress on people's prejudices. Patterns of geographical change rest on *images* that resist or promote changefulness: hence the importance of William Kirk's 'cultural eye' and the way people look at their environment. His emphasis on the value of values in geography has had a great impact and has led to the growth of a value-orientated geography which provides a new key to the understanding of the earth. The patterns of today are the prejudices of the past locked into the fantasies of the future. This is what geography must be more and more about.

Notes

1 J. W. Watson, 'Image geography: the myth of America in the American scene', *Advancement of Science*, vol. 27 (1970), pp. 1–9.
2 J[essie]. W. Watson, 'The teaching of value geography', *Geography*, vol. 62 (1977), p. 198.
3 E. Jones, 'Village, town or settlement system: discussion', in W. D. C. Wright and D. H. Stewart (eds), *The Exploding City* (Edinburgh University Press, Edinburgh, 1979), p. 127.
4 Quoted out of context but not inapplicably from F. M. A. de Voltaire, *The Liberty of the Press*.
5 R. H. Osborne and R. Jones, 'The geographical distribution of professional persons, with special reference to the place-attitudes of intending teachers in Scotland', unpublished paper delivered at the Glasgow meeting of the British Association for the Advancement of Science, 1958.
6 M. S. Samuels, 'The biography of landscape: cause and culpability', in D. W. Meinig (ed.), *The Interpretation of Ordinary Landscapes: Geographical Essays* (Oxford University Press, New York and Oxford, 1979), pp. 51–88.
7 Edinburgh Town Council, *Edinburgh, 1329–1929, the Sexcentenary of the Bruce Charter* (Oliver and Boyd, Edinburgh, 1929), p. 181.
8 D. Wilson, *Memorials of Edinburgh* (Jack, Edinburgh and London, 1886), pp. 172–3.
9 D. Stewart, 'Preface', in A. Smith, *Essays on Philosophical Subjects* (Creech, Edinburgh, 1795 and London, 1811), p. li.
10 A. Smith, *The Wealth of Nations*, ed. J. E. H. Rogers (Clarendon Press, Oxford, 1880), vol. I, book II, ch. III, p. 338.
11 Ibid., vol. I, book II, ch. III, p. 344; vol. I, book I, ch. XI, pp. 263–4.
12 Stewart, 'Preface', in Smith, *Essays on Philosophical Subjects*, p. xxxvi.
13 Ibid., p. xii.
14 Smith, *Essays on Philosophical Subjects*, p. 135.
15 Ibid., p. 124.
16 Old Edinburgh Club, *The Book of the Old Edinburgh Club* (Constables, Edinburgh, 1911), vol. IV, p. 52.
17 A. J. Youngson, *The Making of Classical Edinburgh* (Edinburgh University Press, Edinburgh, 1966), p. 70.
18 Ibid.
19 Smith, *The Wealth of Nations*, vol. II, book V, ch. I, p. 389.
20 Ibid., vol. II, book IV, ch. III, p. 69.
21 Youngson, *The Making of Classical Edinburgh*, p. 10.
22 See J. Appleton, *The Experience of Landscape* (John Wiley, London, 1975).
23 E. Graham, 'Edinburgh 1850–1900 – Victorian city', in J. B. Barclay (ed.), *Looking at Lothian* (Royal Scottish Geographical Society, Edinburgh, 1979), p. 61.
24 G. S. Dunbar, *Elisée Reclus: Historian of Nature* (Archon, Hamden, Conn., 1978), p. 44.

25 P. Boardman, *The Worlds of Patrick Geddes* (Routledge & Kegan Paul, London, 1978), p. 4.
26 P. Kropotkin, *Mutual Aid: A Factor of Evolution* (Heinemann, London, 1902).
27 R. Fishman, *Urban Utopias in the Twentieth Century* (Basic Books, New York, 1977), p. 33.

Chapter seven

Small-town images: evocation, function, and manipulation

Brian Goodey

Small British market or country towns are so commonplace, so normal, that they have avoided the analysis and concern of the geographer and planning theorist. In what follows I want to hint at a reconsideration of these environments which commonplace experience has made almost invisible. My reasons for doing so concern both a general regard for human development – as US poet Charles Olson noted, 'Man is forever estranged to the degree that his stance toward reality disengages him from the familiar'[1] – and a specific concern about the current problems of these towns and their users. I believe that massive changes are affecting the towns and that new geographies, sociologies, and images of place are being manipulated with little public or professional comment.

It must be admitted that 'progress' in both geographical analysis and planning procedures has served to obscure the fundamental issues concerning image and behaviour pioneered by William Kirk. What rapidly but erroneously became known as 'perception studies' were preoccupied with technique and, in analysing spatial structure, evaluation, or preference, fragmented the composite of popular images upon which behaviour in urban space is predicated. Today, only retail location and facility design use much searched-for information in competitive image-making, whilst that traditional perceptual concern, urban legibility, is seldom discussed in planning circles.

It is within the context of the small but growing group of geographers concerned with interpretation of the environment that I set my remarks, incorporating the broad range of images and media for communicating them as did Kirk in those creative developments of his pioneer texts, which were unfortunately seldom reduced to written form.

As to defining 'small town', I share with other writers a difficulty in relating image to population size. I have lived, for example, in a small US Midwestern *town* which is actually smaller than the

111

expanding *village* where I now live. Oxford, on the other hand, is a city in many respects, and yet it maintains those views out to the former agricultural hinterland which often mark the 'country' town in the British mind. In a recent account, Chamberlin suggests that the country town is any place which preserves the memory of fields within town boundaries, but as this admits London, he also notes that this definition is 'ambivalent, contradictory, but, to the English, meaningful'.[2]

My own meanings have been shaped by a succession of studies in which I have sought to reveal popular perceptions of a range of such towns, to endorse their formal and informal functions, and to chart recent changes.[3] My view is that of an insider, a country-town product and user, still driven by public transport to a centre. Not for me the blurred skyline, the faces looking out, seen by Thomas:

> Of Basingstoke in Hampshire
> The claims to fame are small:
> A derelict canal
> And a cream and green Town Hall.
>
> At each week-day the 'locals'
> Line the Market Square,
> And as the traffic passes,
> They stand and stand and stare.[4]

I expect, and usually find, a centre with the predictable array of standard functions – shops, pubs, hotels ... possibly a station – which Betjeman described in his topographer's strategy of the pre-war Shell guide era.[5] From the built form I expect Sharawaggi – 'the art of making urban landscape ... the antithesis of symmetry and neo-classicism' – which Casson explored with Cullen, Bawden, and the rest at the birth of the 'townscape' urban-design movement in Britain.[6]

The popular design language of human scale, complexity, flexibility, and so on, finds its sources in the country town, a source which could, surely, stand firm and adapt to those limited pressures which might touch it. Simmons describes the image of Chipping Campden but speaks also for our expectations of the type:

> The main street of Chipping Campden is one of the most completely harmonious in England; and being English, its harmony is composed of diversity. The buildings in it range from the 15th century to the 20th. There is not one formal terrace, no

group of adjacent houses designed alike. No single building stands out as more important than the rest. The little town as a whole, and above all this street and the approach to the church, are more striking than any individual element in them.[7]

That physical harmony may also indicate a presumed tradition of good human associations is evident in the modest challenges offered by 'Borchester', the country town in which the BBC radio serial 'The Archers' is set. And this is even more clearly painted in North American preoccupations. In his journey along *Blue Highways*, for instance, Trogdon came to Bagley, Minnesota, the small town of my US residence and

... walked down to the bakery, the one with flour sacks for sale in the front window and bowling trophies above the apple turnovers. The people of the northern midlands – the Swedes and Norgies and Danes – apparently hadn't heard about the demise of independent, small-town bakeries; most of their towns had at least one.

With a bag of blueberry tarts, I went up Main to a tin-sided, false-front tavern called Michel's, just down the street from the Cease Funeral Home. The interior was log siding and yellowed knotty pine. In the backroom the Junior Chamber of Commerce talked about potatoes, pulpwood, dairy products, and somebody's broken fishing rod. I stay at the bar.[8]

However, if your bar was in New York or Los Angeles, the best-selling radio series-cum-fiction of Lake Wobegon could continue this very personal journey to a harmonious heartland of just yesterday.[9]

Comparison of US small towns and the British country town is, of course, dangerous – the geography is so different. US small towns are allowed to exist, boost, stumble, or fall in rural stretches which will always be only a benign fiction to many city dwellers. Academically, they can be looked at and left safe in the knowledge that only some will be drastically modified by the next generation. Nevertheless, McClung, a Professor of English writing in an architectural journal, has formed a proposition from Lynch's work which, I believe, deserves consideration in the British context:

... continuity and openness, individual growth and collective survival. The idea of the small town is probably fated always to exist within a matrix containing these opposing and at times self-destructive forces; its success – that of the town, and of the idea of the town – lies in striking the balance between them.[10]

The presence or absence of these human values in British small towns has seldom been questioned. As Weightman noted, 'Small towns anywhere suffer from a kind of metropolitan prejudice. They are backwaters, claustrophobic enclaves for those without ambition; they lack both the rural idealism of the village and the excitement generated by big cities.'[11] Aside from a short burst of 'community' study, students and academics have ignored them, but since the wartime analysis of a future Britain, planning interests have been hell-bent on changing or even 'saving' them.[12] The past 50 years of planning studies and actions reveal regional plans, expansion schemes, central-area redevelopments, conservation strategies, retail schemes, and residential expansions – fashions applied to the physical fabric of small market centres. Many have succeeded in their own, limited, terms but many more have failed to tap the ambition, and image, held by the local community, as Nairn noted:

> Nowhere is ordinary. There is always something to build on, some idea from the past or present character to understand or expand. We accept this easily enough when it is applied to individual persons; we ought to be much more ready to apply it to towns, collective persons. Planning all too often produces the lowest-common-denominator when it ought to be aiming for the highest-common-multiple.[13]

Those small-town images which owe least to spatial form and environmental design would seem to endure down the generations – towns are still known locally as 'lazy', 'rowdy' or 'closed' – but there are no agents charged directly with responsibility for such epithets save, perhaps, the local newspaper. When it comes to the conscious management of spatial and built form, the situation is different.

Seldom has there been the continuity in political and professional attention to a town to ensure that key features of the built image – nodes, landmarks, edges – are managed so as to balance continuity of town experience with the necessary novelty of new environmental details which each successive generation can call its own. In the brief period of the 1960s, when perception studies were part of a new, participatory approach to planning it became clear that popular images of place were difficult to reconcile with the place-making materials at the planner's and architect's command. It was also evident that different user-groups held different, overlapping, images of a town: in short, different groups operated within different behavioural environments. Whilst this revealed a richness in even the unexceptional, such evidence often exceeded the time allocation given for public participation in the planning process.

Nevertheless, these overlapping images and patterns of behaviour are the essence of life in the small town. We are long overdue an evaluation of the changes effected in, and proposed for, small country and market towns. So far as the built form is concerned, the stereotypical sequence from sleepy agricultural centre to townscape dominated by shopping blocks can be imagined. But what of the changing human experience, and perceptions of such places?

Some 10 years ago, I wrote that 'going to town' was once a rich experience which regularly endorsed one's decision to reside in a place by offering novelty, tradition, and a context to be shared with fellow citizens. In designing viable commercial centres this human dimension has been ignored. The next decade offers both the opportunity and the necessity for developing the sense of urban place.[14] As I will suggest towards the end, I am not convinced that this sense of urban place has been developed anew, although the main justification for the professional activities of the urban designer has been the enhancement of urban experience, seen largely in the design and development of urban places which encourage positive response from and activities by the many different individuals and groups drawn to them. This function of the city centre or, more commonly, town centre as a place for meeting, watching, precipitating change, and evaluating fashion has been endorsed throughout history. Titled places, typically open spaces, have become linked to both named and dated events, and to the ill-defined traditions and characters that have populated them. Of these events, the seasonal fair and weekly market usually provide the most obvious link between place and space, and the market, at least, retains a significant role in the life of British small towns.

Although the 'market' prefix is increasingly neglected, the market or country town maintains elements of a human and spatial character which derive from those functions described by Rose:

> Going to market always brought with it a spirit of hilarity, as well it might, for a good deal of real drudgery belonged to village life. To put on clean clothes and lighter boots was itself an aid and inspiration to the spirit. Besides, however much one loved the village and enjoyed its life, to get away from it once a week for a few hours made it seem all the pleasanter on the return. And so the weekly market, though it was fundamentally a business matter, had a character beyond mere

115

buying and selling: it was also a good-humoured gathering and meeting together, everybody expected this and responded to it heartily.[15]

On a seasonal basis the market became a fair for hiring and on a weekly basis it was the one contact which all rural residents could have with their fellows, exchanging both goods and information in a special environment. Wait at any country town bus station on market day and something of Rose's pre-war atmosphere can still be sensed as an older generation emerges from the weekly bus.[16] However, although stalled markets for the sale of foodstuffs and household goods have taken on a new lease of life in the past decade, they seldom serve their former purpose. Livestock sales have been increasingly centralized at sale rings in a few towns where hygiene and the universal need for road access have usually led to improved peripheral locations. Behind closed doors, stock sales represent business transactions which are now part of a private world which few town users are able to understand or share.

However, as the 'going to town' experience moved rapidly from its agricultural origins in the 1950s and 1960s it retained a particular significance for at least five groups within the population, each making their peculiar spatial and temporal bids for town and place in the post-war era.[17]

The *housewife*, with limited weekly expenditure and without home refrigeration, developed the almost daily visit to the shops as an essential social, as well as economic, experience. For the unwaged housewife, the patterned exploration of part of the town not only permitted comparison shopping for availability, quality, and price, but also offered the chance to encounter others of like purpose and problems.

The visit to town which provided essential social contact beyond the 'backyard fence' was a major method of information diffusion, and a source of some of the largely informal learning essential to the housewife's role. This latter became all the more evident with child-rearing, with children wheeled daily in their prams, and their development compared. Thus, the daily one- or two-hour trip both for shopping and the use of public amenity spaces was of key significance for this substantial element of the urban and suburban population.

For the *adolescent* the town provided a very different range of opportunities. In the era of single-sex secondary schools, the journey home offered one of the regular opportunities for initial contact with the opposite sex. Bus and rail stations, libraries, and other public places were also pressed into service as viewing galleries. Scribbled notes and frantic messages served to promote opportunities which might be fulfilled on Saturdays, when 'going to town' became the major experience of the teenage week. With little save time to spend, single-sex groups, organized by age, school, or place of origin, circled town centres stopping at record and clothes shops, and lingered in café or coffee bar, in the hope of enlarging their social circles or of improving knowledge of the latest associations. With advancing teenage, pubs and Saturday night dance halls began to enlarge the range of town places occupied.

For the *nuclear family*, Saturday afternoon was, for a time at least, the occasion for joint purchases of other than basic goods. Fathers may well have worked on Saturday mornings, but in the afternoon there was the chance, or duty, to dress in intermediate finery (though not Sunday best) and parade with the family in search of a single item allowed by the family budget. Decisions on children's clothes into early teenage were a family affair, and new items of furniture or household equipment often required several weeks of prior comparison shopping, seldom extended though it was to neighbouring town centres. As for most groups the market region defined by public transport patterns was all-important. There was the chance to make contact with other families, parents meeting old school-friends, and chancing upon the parents of their children's friends. Groups met which assessed the state of the town, and which made casual assertions on the state of the nation. For father this was often the only opportunity to share a world which his wife knew only too well.

For the *elderly*, and especially the widow, the experience of 'going to town' was essential. This was because limited incomes meant that comparison shopping was mandatory, because the ability to undertake the journey was a crucial personal measure of mobility maintained and because social networks, which may formerly have been town-wide through the work or leisure re-lationships of the deceased partner, could only be maintained by the widowed through encounters in the town centre. Various types of tea shops, and quietly defined seating areas, catered for this particlar market, allowing social groups to develop on neutral territory long before voluntary and other agencies became con-cerned with the needs of the elderly.

Finally, we can identify a fifth group – variously described as *individuals, characters, or outcasts* – who found in the experience of 'going to town', or being in town, something of the social contact which was so evidently enjoyed by others who came from, and returned to, their own homes. Post-war evocations of market-town life often identify these characters who occupied, or in some cases clung to, the margins of urban space, either by offering peripheral services in a very public way – newspaper selling, scavenging, market-day cattle droving – or by simply marking public places with their presence. In the insensitive past the classification of 'social inadequates' served to embrace those with mental, physical, linguistic, ethnic, or other differences which marked them as apart from the local population. Their need for casual and neutral spaces was thus made all the more significant.

These town users may be but a memory, an image of the 1950s, which has been gradually eroded by a succession of changes in the subsequent 30 years. But if this is the case, then who 'goes to town' today, and for what purposes?

So much of the seductive, but ephemeral, justification for new urban-design schemes relies on the smiling faces of stick/stock figures that are added to the drawings of public spaces surrounding new retail and commercial proposals. Thus, we need to ask why in reality these figures might be there? The rationale for change in the design, provisioning and access to town centres is now based on the successful market relationship, where 'market' implies public demand for what the developer chooses to provide. It is therefore appropriate to examine some of the social and cultural meanings ascribed to town centres and to ask if urban designers are reflecting them, as well as the more insistent demands of commercial, institutional, or local-authority clients.

Changes in personal circumstances and urban structure have gone hand in hand over the past 30 years and there is certainly no implication in what follows that terrible things have been done to the town user, or at least to the majority of town users, without their willing acquiescence.

The rapid rise in car ownership, from family weekend luxury to everyday runabout, was promoted by the car industry, but the mobility which followed was seldom seen as other than a positive asset to both family and community. The first High Street super-markets, offering reduced prices through bulk wholesale purchase, the combination of several speciality foodstores under one roof, were likewise seen as an asset by most shoppers. Self-service, access to standardized products 'as advertised nationally', covered shopping centres, and pedestrianized environments have all had obvious personal benefits and there has been little orchestrated public discussion of the disbenefits which might result.

For a considerable period, however, there was a substantial barrier to retail innovation in Britain. The planning system was set against the development of the subregional, out-of-town shopping centre which had existed in the United States since the 1930s and which was spreading through the remainder of Western Europe in the 1960s. Until the mid-1970s no such centre appeared on the British landscape without there being a hard-fought battle where the inevitable self-interest of existing town-centre retailers and landowners was supported by official judgement as to the significance of the town centre as the spatial and social focus of urban life. Gradually, however, warehouse-like retail outlets on route intersections led the way in showing that the mobile family was only too pleased to embrace one-stop shopping. In the past few years the challenge of network intersection facilities has grown rapidly with little public debate as to the quality of experience in either the new, or the old, retail areas. The names of towns have been replaced by those of superstore retailers in many personal geographies.[18]

Each of the groups described in the 1950s has had to adjust to major changes in the past 30 years. Even the term *housewife* has become unacceptable in a literature of social change which has charted the move from the 1950s to the 1980s, although it is clear that many women still describe themselves as such, or are preoccupied by those earlier basic concerns as a matter of necessity. The daily shopping visit has been affected by a number of factors, notably increased employment of women and the growth of the one-stop weekly, or fortnightly, shopping trip by car. Refrigeration and the growth in pre-packaged, preserved, convenience foods, the replacement of weekly wage packets by salary cheques, and access to the family car have all reduced the need for the daily visit.

Personal habits, limited incomes, short-term domestic planning and the continuing need for social contact all suggest that many home-based women do continue to make the visit to town. For those on low incomes the period between incoming cash and expenditure is inevitably brief, and, for those at higher income levels, comparison shopping for supplementary or specific items (such as meat, fish, or bread) can still justify frequent town visits, the revived stalled market often providing a target for both income groups. The chance for social contact and for child comparison still persists, even if the stately pram has been replaced by a back seat in the car and a more modest push-chair.

For the *adolescent* (another term in declining usage) life has probably changed less than for the other groups identified. Although often described by their elders as being 'not what they were in our day', teenagers and their needs in the town centre seem to have changed very little in 30 years. True, there are far fewer inter-mediate zones offering sheltered semi-public/semi-private space where the initial courting overtures can be rehearsed, though the need has not declined! Even aside from the special case of football supporters, knots of 'youths' still appear to threaten others by their mere presence, and by their failure to imply a sufficiently commercial purpose in environments which are increasingly designed with effective sales in mind. Whilst the language of youthful display may be different from that of the past, behaviour is probably no more eccentric than it was in the context of the 1950s town, when there appeared to be rather more tolerance on the part of authority. Today, commercial invitations are more insistent and more pointed and the teenager who can only stand and stare is made to feel that he or she should have the wherewithal to do more, to buy into the much vaunted music and clothing worlds which are at the same time securely guarded by private, if not public, police.

Saturday visits by the *nuclear family* have probably seen the biggest change over the past thirty years. Originally there was a need for the family to be seen as a family and as part of a close-knit community. Population mobility has meant that there is less chance to see, and be seen, by long-standing family connections. The family is a lot less likely to survive as a unit, and if it does so it is more likely to be packaged in a car. Father and mother are now more likely to have working hours which allow at least some evening shopping for the larger household items, where, in any case, there is less likely to be the need for the father to show a public face of 'providing for' the family. Do-it-yourself repairs, car care, Saturday afternoon sports TV, a much wider range of leisure opportunities, car mobility, and the inherent dislike by many men (and not a few women and children) of being dragged around crowded shops to view goods which are neither needed nor affordable, are other factors which make the young couple without children the most active shoppers in the Saturday afternoon townscape.

Although there have been some modest changes in provison for the *elderly* such as day centres, clubs, and special transport, it is clear that the social importance of 'going to town' has not declined over the period under discussion, even if the necessity for shop-ping has been reduced. Regular visits to the Post Office to collect pensions have long been established as ritual encounters. Just as the teenager seeks to impress a new image on the broader society which the town centre represents, so the elderly see their visits as the major opportunity to confirm that they are still part of society, a society which is continually endorsed as 'normal' through both the fictions and commercial invitations of television. Towns retain, as nowhere else, evidence of good times past and, although there is much to complain about, the ability to function regularly as on-the-spot commentator should not be undervalued. Prices, the privatization of casual resting places, parking difficulties, and reduction in public transport may all serve to deter the elderly town user.

For the *character* the town still provides a place of validation, security, and protest. Looking back to the outsiders of the 1950s there seems to be little continuity; caring agencies have swept earlier definitions of the inadequate from the streets. Such has been the recent (though hardly current) belief in a caring society that we now see the outsiders, including the larger number of unemployed, that remain as those who do not want to be helped. Indeed, this is quite likely to be the case, inasmuch as they want to use the spaces, casual warmth, dynamism, and available audiences in an environment which they are forced to design for themselves. Once they, and their predecessors, were tolerated as characters, but are no longer regarded as appropriate to the more organized patterns of town-centre life.

The drunk and the punk, trying hard to be outsiders, are monitored and controlled as unfortunate problems in modern town centres, their life-styles proving unadaptable to what is provided. For the rest, housewife, adolescent, the family, and the elderly, for all on the fringes of the predominant commercial target group of young upwardly mobile spenders, adaptability is expected and, however reluctantly, occurs.[19].

A people- rather than consumer-centred view of contemporary town use has received little consideration in discussions of town-centre design. Structured behavioural analyses of public space, such as that by Korosec-Serfaty,[20] are rare and seem seldom used by designers or planners, who prefer the blunt instruments of commercial shopping surveys. Although behavioural work remains an essential element in US architectural schools, and therefore stands a modest chance of influencing community design practice, it is seldom evident in British studies. Design practice in Britain relies increasingly on 'off the peg' imagery for the detailing of buildings and spaces within planning briefs which seldom imply expected or desired patterns of behaviour, save access to retail, office, and the occasional residential property. The tradition of both critical and behavioural observation in urban design has declined in the face of economic-development imperatives and, as a result, some essential market-town characteristics require restatement.

Application

The illustrations below (Figures 7.1–7.10)[21] should serve as a reminder to those aware of Cullen's *Townscape* approach which long dominated British urban-design thinking.[22] These examples are largely drawn from Llandeilo, a small market town, and Kettering, a significant regional centre, but the themes are evident at every scale of centre and each pair of photographs highlights an issue which deserves further exploration.

Traditional ideas on *entry and exit* (Figures 7.1 and 7.2) suggest that the town should have a clear 'start' or 'edge' signified by gateway features. How significant is this characteristic in defining a place worthy of regular visits, and how have the free-standing gates of shopping centres affected this imagery? Beyond this the importance of *street rhythm* in characterizing traditional market towns has also been stressed, this being a morphologically derived device to be maintained through strict development control and attention to elevation detail (Figure 7.3). But when the 'rhythm' (which is susceptible to analysis) is broken (as in Figure 7.4) is anything more than a local aesthetic rule offended? Can the patterns of pedestrian use in traditional and more modern streets be assessed, and what of the more subtle responses by individuals and groups to the changing townscape over time?

I would not go as far as Brand[23] in suggesting that the 'sense of place' becomes dis-located (locus free) and defined instead as a consumer item, but the stalled markets survival (Figures 7.5 and 7.6) and growth is a positive reminder of the original small-town function which may fit in with Brand's view. Chartered markets expand, Sunday markets and their poor relations, the car-boot (trunk) sales, grow in peripheral locations as the uncovered and unpoliced antithesis of the new shopping centre. There are no consumer or property surveys here, just another element in popular culture which evades the planner and is unrecognized by the researcher.

Scale and pace are shown with negative imagery from Kettering (Figures 7.7 and 7.8) but here the search must be for degrees of tolerance, rather than absolutes. The universality of 'sites awaiting development', of piecemeal infill, of shop façades hiding the death of a building is accepted as the price which a town centre pays for change, or 'development'. But how much – in terms of both scale and pace – of this interrupted 'progress' can town users take before rejecting the town and turning their interests elsewhere? The content of Figure 7.7 largely results from the influence of distant financial institutions, that of Figure 7.8 from the more local demands of security-conscious shop-owners and maintenance staff. The latter photograph, taken at 5.45 pm on a weekday, presents

Figure 7.1 Llandeilo, Dyfed: South-east to the Tywi valley across the churchyard of St Teilo's, which straddles the A483.

Figure 7.2 Llandeilo, Dyfed: North-east from the same point, daily commercial and tourist traffic fight through the 'gate' of two banks which hint at the town's status as county seat of Carmarthenshire until the end of the nineteenth century.

Figure 7.3 Llandeilo, Dyfed: Carmarthen Street looking to its junction with the A40 at Rhosmaen Street. Variable street width, snaked foreclosure of view, and stepped, human-scale façades of shops and homes provide optimum interest in short distance.

Figure 7.4 Kettering, Northants: High Street with another bank 'gate' from Market Place. Larger town, larger scale, but Midland Bank and Bank Chambers to the left still admit 1930s Granada cinema as a neighbour. View now open to horizontal recent development which removes transition between commercial and retail areas.

Figure 7.5 Llandeilo, Dyfed: Modest and informal market, relocated on path between retail and commercial 'high street' on Rhosmaen Street and a large backland car park with tourist information centre nearby. Considerable unpressured sitting and standing space remains.

Figure 7.6 Corby, Northants: Large formal market in repaved square, backed in this view by the pre-war, company town centre buildings. A third of the stalls are devoted to bric-à-brac of the 'car boot/trunk' type.

Application

Figure 7.7 Kettering, Northants: 'Anywhere' view from the rear of new town centre developments, over rough waste car park 'awaiting development' to hoarding, bus bays, and turn of the century commercial buildings with unsympathetic ground-floor shop façades . . . and neighbour. The projected pace of change is unfulfilled, the town's scale is shattered.

Figure 7.8 Kettering, Northants: The eighteenth-century Wadcroft field name does little to convince that this view of old and new, seamlessly welded by wall-to-wall floorscape, represents continuity with local tradition. The informality of urban streets has been pedestrianized for easy maintenance and security.

an environment so sterile that the security doors might as well be closed across the street itself.

Finally, we turn to what might seem to be the fine *detail in public space* (Figures 7.9 and 7.10) but at a scale where the shifting line between public and private space is officially defined and unofficially redefined.

The non-purchasing public has been recognized as being on the retreat since the 1970s but what should the town contain besides shopping opportunities, and how are the various users to be catered for? One fairly obvious answer is to ask them, and to work with them in defining and deciding on appropriate design solutions for the market-town environment.

Reviewing imposed 'big city' planning solutions as applied to small towns in North America, Cohen and Hodge[24] have pointed out that if participatory or community design is to work anywhere then it must be in the small town. The projects developed by Sanoff[25] and techniques presented in the journal *Small Town* are illustrative of the methods adopted in the USA in the mid-1970s. These US small towns are generally smaller than the market towns on which we are focusing here, but the success of the 1974 *Vivat Ware!* report for that Hertfordshire market town on the London fringe shows that an open and provocative consultant's report can stimulate wide local response.[26] The Civic Trust's model scheme for small-town regeneration in Wirksworth, the more diverse initiatives of the local community in Macclesfield, the nearby Pennine Heritage Network of communities, and the more recent Calne Project near Bristol all provide examples of market-town regeneration, usually with a strong historical and tourist theme.[27]

The success of community-based regeneration initiatives which revive or invent a market for historical townscapes complicate rather than resolve the future of the market town. For now we must admit to the existence of at least three different 'small-town' central environments, each with increasingly specific access, design, and promotional demands, as well as its own regional network of competitors. First is the successful market centre where new shopping centres seem to have been grafted on to restored traditional streets and parking has replaced in-town agricultural industry. Second is the out-of-town shopping centre, a new location and a new name on the map, with each successive addition containing more peripheral leisure attractions to capture the car-loose shopper. Third is the well-placed tidied townscape, marketing its industrial past as craft and providing an array of speciality shops to replace basic suppliers who are no longer required. Each 'small-town' type requires specific design skills and

Figure 7.9 Kettering, Northants: Entrance to shop in an area subject to redevelopment. Between tight pavement and small shop units, this public/private transition is marked by commercially effective design detail.

Figure 7.10 Kettering, Northants: 'Planter' in paved pedestrian area. Here the public/private space divide is clearly marked by streetline shop façades, the informal public space is 'provided' rather than evolved element in the town's morphology, and responsibility has faded.

is evolving a different townscape appropriate to its economic needs. There is, of course, a fourth type of town. Here, and probably still in the majority, are those towns of the sleepy shires as well as the industrial North, where no new image or purpose has been proposed, where a premature shopping centre decays, where listed buildings list, where job opportunities decline, and where planners are unable to attract development finance or grants.

Whilst I accept that the traditional British market town may inevitably be fragmenting I have only a slight view of what this will mean to town users. In part this meaning may be deduced from the readily available accounts of consumer spending, coupled, perhaps, with police reports as negative evidence. What of the informal functions of town space, of special days, of patterned walks and visits, of chance and planned encounters in predictable places? Will traditional townscape criteria for design be appropriate only in the historic or Disney-derived 'fun' town, and if so what are the new behavioural and aesthetic criteria for the alternative varieties? Will any opportunities remain for the temporary structuring of public space by casual users, or will the stick/stock people take over? I think we ought to be told, and preferably by a new generation of geographers who recognize the importance of quality in everyday contemporary experience.

Notes

1 C. Olson, *The Special View of History*, ed. A. Charters (Oyez Press, Berkeley, 1970), p. 29. Olson, founder of the Black Mountain school of poetry, devoted much of his attention to Gloucester, Mass.; I have versions of this quotation in an exchange of letters between us, but this appears in R. Melnick, 'The regional character of towns: notes from the flint hills of Kansas', *Small Town*, vol. 8, no. 6 (1977), pp. 4–10.

2 R. Chamberlin, *The English Country Town* (Webb and Bower/National Trust, Exeter, 1983); the question of definition is raised by A. Clifton-Taylor in his review, 'Talk of the town', *Architects' Journal*, vol. 178, no. 38 (21 September 1983), p. 99.

3 Towns from my home region are explored in *East Anglia's Built Environment as an Education Resource* (East Region, RIBA, London, 1981). The significance of human associations rather than built-form imagery is discussed in 'People, places and/or things . . .', *Bulletin for Environmental Education*, no. 138 (November 1982), pp. 22–3. For a comparison between a town as formally presented to visitors and as experienced see 'Values in place: interpretations and implications from Bedford', in J. R. Gold and J. Burgess (eds), *Valued Environments* (Allen & Unwin, London, 1982), pp. 10–34. Origins in a large country town, Chelmsford, emerge in 'Lynton Lamb's "Country Town" revisited: Chelmsford, Essex', paper to 'Geography

and Literature: Experience of Place' session, Institute of British Geographers, Durham, 1984; and, more recently, in B. Goodey, L. Smales, C. Bray, and P. Thompson, *A Field Guide to Urban Design in Essex* (Joint Centre for Urban Design, Oxford Polytechnic, Oxford, 1989). A major influence has been a five-year project with the Council of Europe, reported in B. Goodey, *Urban Culture at a Turning Point: A Council of Europe Project in Twenty One Towns* (Council of Europe, Strasbourg, 1983).

4 'Basingstoke' by L. Thomas appears in G. R. Hamilton and J. Arlott (eds), *Landmarks* (Cambridge University Press, Cambridge, 1943), p. 56.

5 J. Betjeman, *English Cities and Small Towns* (Collins, London, 1943): the book begins with a discussion as to how to experience a small town based on the work by John Betjeman and John Piper for the early Shell guides.

6 H. Casson, 'The obliging Sharawag', in A. G. Weidenfeld and H. de C. Hastings (eds), *Grand Perspective: A Contact Book* (Contact Publications, London, 1947), p. 65.

7 J. Simmons, *A Selective Guide to England* (John Bartholomew, Edinburgh, 1979), p. 122.

8 W. Trogdon (writing as William Least Heat Moon), *Blue Highways: A Journey into America* (Fawcett Crest, New York, 1982), p. 291. Even Trogdon calls Bagley a village, but to its residents and region it demanded the legal and experiential respect of a town in the 1960s, though depopulation and location may serve to demote it for successive generations.

9 G. Keillor, *Lake Wobegon Days* (Penguin, New York, 1986). In reviewing this book, based on a ten-year run of *radio* programmes, J. D. Brown of the *Chicago Tribune* noted, 'a comic anatomy of what is small and ordinary and therefore potentially profound and universal in American life. . . . Keillor's great strength as a writer is to make the ordinary extraordinary'.

10 W. A. McClung, 'The mediating structure of the small town', *Journal of Architectural Education*, vol. 38, no. 3 (1985), pp. 2–7. McClung draws from Kevin Lynch's *A Theory of Good City Form* (MIT Press, Cambridge, Mass., 1981). On the US small town see, for initiation, C. Rifkind, *Main Street: The Face of Urban America* (Harper & Row, New York, 1977); D. M. Humman, 'Popular images of the American small town,' *Landscape*, vol. 254, no. 2 (1980), pp. 3–9; J. A. Jakle, *The American Small Town: Twentieth-century Place Images* (Archon, Hamden, Conn., 1982.

11 G. Weightman, 'The small town life', *New Society*, vol. 31, no. 10 (1974), p. 265.

12 The future role of the country town is explored in detail by planners and politicians in S. Baron (ed.), *Country Towns in the Future England: A Report of the Country Towns' Conference of the Town and Country Planning Association* (Faber & Faber, London, 1944), where the social benefits of 'overspilling' city populations to small towns are

discussed; my nearby town of Brackley, Northants is featured and its 'potential', though unrealized in favour of Daventry, is further detailed by M. Gregory, 'Declining population: possible remedies in South Northants', *Town and Country Planning*, vol. 20, no. 3 (1952), pp. 119–25.

13 I. Nairn, 'A fanatic's guide to the great citadels of football', *Sunday Times Weekly Review*, vol. 12, no. 5 (1974), pp. 33–4.

14 B. Goodey, 'Going to town in the 1980s: towards a more human experience of commercial space', *Built Environment*, vol. 5, no. 1 (1979), pp. 27–36.

15 W. Rose, *Good Neighbours: Some Recollections of an English Village and its People* (Cambridge University Press, Cambridge, 1942), pp. 82–3.

16 The 'bus to market' has been promoted as a tourist attraction; meanwhile bus deregulation and the marginalization of bus stands in many medium-sized towns has removed the bus station as a central meeting place. For the tourist market see E. Gundry, *To Market by Bus* (map) (Public Affairs Department, National Bus Company, London, 1983) and R. Anderson, *The Markets and Fairs of England: A Buyer's and Browser's Guide* (Bell & Hyman, London, 1985), which contains a useful current market listing.

17 This commentary has benefited from the experience of students in an 'Historic and Tourist Environments' option. A version was circulated as 'Going to town: the popular experience of market towns and their design', Research Note 20 (1986), Joint Centre for Urban Design, Oxford Polytechnic, Oxford.

18 The past 15 years of retail evolution have been described in a large academic and corporate literature. Still of interest as herald of the inevitable is the National Economic Development Council's *The Future Pattern of Shopping* (HMSO, London, 1971). See also V.J. Bunce, 'Revolution in the High Street? The emergence of the enclosed shopping centre', *Geography*, vol. 68, no. 301 (1983), pp. 307–18, and, with an appraisal of current issues, R. Schiller, 'Retail decentralisation – The coming of the third wave', *Planner*, vol. 72, no. 7 (1986), pp. 13–15, based on research by Hillier Parker, and W. King, 'The future roles of town centres', *Planner*, vol. 73, no. 4 (1987), pp. 18–22.

19 'There ought to be ways of making places safer other than by increasing policing and enjoining people to stay behind closed doors. Pleasurable though it is, there must be more to life than shopping', concludes K. Worpole, 'Part-time places', *Guardian*, 16 September 1987, in a review essay based on the South-East Economic Development Strategy reports *On the Town* and *Trade Winds* published in 1987.

20 P. Korosec-Serfaty, 'The main square: functions and daily uses of Stortorget, Malmo', Aris Nova Series 1 (1982), University Art Institute, Lund employs a multi-faceted approach to the meaning, design, and use of an urban space but retains a site, rather than a user focus.

21 The photographs of Llandeilo date from August 1986, those of Kettering and Corby from September 1987: it is possible to provide a negative view of Llandeilo and a positive image of Kettering!

22 G. Cullen, *Townscape* (Architectural Press, London, 1961, but subsequently 1974 in an abbreviated form). Cullen's work has frequently been criticized for the absence of theoretical underpinning, or clear behavioural analysis. It has, however, served as a key element in design strategies which have incorporated social and behavioural elements, especially with regard to the perception of public and private space as, for example, in J. F. Barker, M. W. Fazio, and H. Hildebrandt's *The Small Town as an Art Object* (School of Architecture, Mississippi State University, Starkville, 1975) and J. F. Barker, M. J. Buono, and H. Hildebrandt, *The Small Town Designbook* (Center for Small Town Research and Design, School of Architecture, Mississippi State University, Starkville, 1981).

23 P. Brand, '"What are you doing here?" asked Milligan or the physics and metaphysics of town centres', *Planner*, vol. 73, no. 4 (1987), pp. 23–6, a cavalier approach to fun-retailing and serious town centres.

24 R. A. Cohen, 'Small town revitalization planning: case studies and a critique', *American Institute of Planners Journal*, vol. 43, no. 1 (1977), pp. 3–12; G. Hodge, 'The citification of small towns: a challenge to planning', *Plan Canada*, vol. 21, no. 2 (1981), pp. 43–7.

25 H. Sanoff and others, *Small Town Design Resourcebook for Small Communities* (Community Development Group, School of Design, North Carolina State University, Raleigh, 1984), has some material in common with A. S. Denman (ed.), 'Design resourcebook for small communities', *Small Town*, vol. 12, no. 3 (1981).

26 R. Townsend, G. Cullen, and A. Henderson, *Vivat Ware!: Strategies to Enhance an Historic Centre* (East Hertfordshire District Council, Ware, 1974). The results, consultants, and community are constructively revisited by G. Darley, 'Town treatment', *Building Design*, vol. 25, no. 10 (1985), pp. 32 et seq.

27 See H. Teggin, 'New life for Wirksworth', *Architects' Journal*, vol. 177, no. 23 (8 June 1983), pp. 48 et seq., and R. Cowan, 'Calne: is there life after sausages?', *Architects' Journal*, vol. 184, no. 51–2 (17–24 December 1986), p. 13. The experience of small-town regeneration is spreading rapidly in Europe, with exchange of examples, as in Council of Europe, 'Conservation policies and urban management in small and medium-sized towns', Urban Renaissance in Europe Study Series 32 (1987), Council of Europe, Strasbourg.

Chapter eight

Personal Construct Theory, residential decision-making, and the behavioural environment

T. J. Anderson

Traditional geographical approaches have viewed human beings more as objects than subjects, victims of circumstance rather than controllers of destiny, pawns rather than players. The behavioural approach looks to the subjective world of perceptions, feelings, attitudes, and beliefs for its contribution to the central aim of geography – the explanation of spatial behaviour. It has been hard to establish the relationship between spatial behaviour and this complex mental world, and it is the purpose of this chapter to show that the concept of decision-making helps to forge that link.

Building on his 1952 paper (reprinted as Chapter 2 in this volume), William Kirk later developed more fully the ideas of the behavioural and phenomenal environments, applying them to the whole of geography and not only to historical research.[1] He identified decision-making as 'one of the most important categories of problems implicit in the nature of the Geographical Environment'.[2] In his model of the decision-maker's environment (see Figure 8.1),

> the social and physical facts of the Phenomenal Environment are shown to constitute parts of the Behavioural Environment of a decision-taker (D) only after they have penetrated a highly selective cultural filter of values. . . . The Behavioural Environment is thus a psycho-physical field in which phenomenal facts are arranged into patterns or structures (*gestalten*) and acquire values in cultural contexts. It is the environment in which rational human behaviour begins and decisions are taken which may or may not be translated into overt action in the Phenomenal Environment.[3]

In order to operationalize the behavioural environment concept we need a sound measurement technique. We are, however, still short of the tools to measure the behavioural environment in the fullness of Kirk's conceptualization, particularly in regard to decision-making.

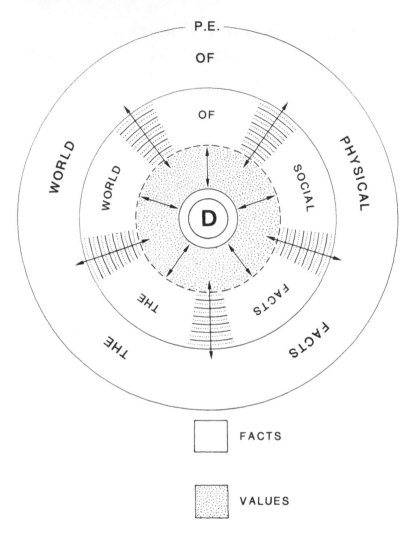

Figure 8.1 The behavioural environment of a decision-maker

Source: William Kirk, 'Problems of geography', *Geography*, vol. 18 (1963), p. 367, table 6

The environmental image work is potentially valuable for this purpose, but on its own falls short because it focuses on the individual more as a perceiver than a decision-maker. The focus has thus been more descriptive than explanatory. Many theories have been developed, however, to explain spatial behaviour in terms of human motivation and nowhere more so than in the area of residential studies. What is lacking among these theories, all of which are plausible and intuitively acceptable, is some means to assess which decision-making factors predominate under certain circumstances. In this chapter environmental images and existing theories are brought together in the hope of throwing fresh light on the nature both of decision-making and of environmental images.

By making two assumptions about the environmental image we will be in a position to detect the presence or absence of certain subprocesses which correspond to the various theories and concepts of decision-making. The first assumption, implicit in most work in behavioural geography, is that the decision-maker's image of his or her environment is vitally involved in his or her decision-making. The second is that it is possible to trace the impact of the past and present decision-making factors in the contours of that image.[4]

Repertory Grid methodology and Personal Construct Theory

The method used in this study to measure the environmental image is the Repertory Grid (RG) technique. It is designed both to record which objects or aspects of the environment are perceived and evaluated, and to reveal something of how the respondent structures this view. Things, places, or people, which are referred to as grid 'elements', are evaluated in terms of 'constructs' (operationalized by attitude scales), along which the respondent can perceive variation. While the individual responses may be of interest in isolation, a principal components analysis of the grid can reveal the major dimensions which the respondent uses for conceptualizing the items under consideration. This corresponds with the *gestalten* mentioned above. What sets the RG technique apart from other measurement tools, such as questionnaires, check-lists, and pictorial displays, is not just analytical sophistication, but above all its foundation on a well-integrated psychological basis, Personal Construct Theory (PCT).

PCT, and the associated RG technique, were devised by a clinical psychologist, George A. Kelly.[5] His aim was to uncover the detailed structure of an individual's perception of part of his or her environment. The single most important concept in PCT is the *construct*, along which the individual can discriminate among elements. Constructs are bipolar, such as good–bad or noisy–quiet. The RG technique is simply a grid which allows each element to be rated on each construct using some scoring technique – seven-point semantic differential scales were used in this study. According to PCT, the individual's image of reality corresponds to a distribution of elements within a reference space defined by constructs; the elements may be visualized as points located along axes which are defined by constructs. This dimensional approach (implemented in the INGRID computer program)[6] provides parsimonious descriptions of the complex grid data and also allows diagrammatic representation which eases the task of the interpretation.

Use of RG technique in geography has been primarily as a descriptive tool. Too often the results have been inspected in the hope of suggesting a hypothesis. Rarely have any hypotheses been used, explicitly at least, to predict the structure or content of the image. By using established theories in the field of residential studies it should be possible to predict at least some of the main features of a typical image from a specific context. By comparing such hypothesized images with the observed images, the relevance or otherwise of the various concepts and theories should become apparent.

Consensus grids

Owing to the clinical origins of RG technique, some significant modifications are required to adapt it to fieldwork conditions, and this brings attendant difficulties of analysis and interpretation. The main modifications relate to the psychologist's concern with the unique mental image of specific individuals, and the way in which the image changes during a course of treatment. This is in contrast to the geographer's interest, which is usually orientated towards the common features of images held by a group at just one point in time. While the psychologist can afford to spend many sessions, each of a few hours length, with his or her client, the field-worker will inevitably be much less involved with a respondent. The most important difference, then, is that one stage is often omitted in the environmental research context, that in which the respondent's own constructs are painstakingly elicited to complete a fully personal grid.

In the so-called 'consensus grid', developed by Slater,[7] individual Repertory Grids must contain not only the same elements but also the same constructs. Both elements and constructs must therefore be supplied by the researcher. The consensus grid is formed by the means for each element expressed as deviations from the general mean for each construct. Because the consensus grid will highlight what people hold in common, and inevitably removes sensitivity to individual variations, the study group should be as homogeneous as possible. For analysis, the consensus grid is treated as if it were an ordinary Repertory Grid.

Since decisions about residential mobility are the subject of this chapter, the images studied focus on aspects of housing. Constructs and elements to be included in the RG were initially selected from a wide range of social-research literature. Since there was no *a priori* reason to consider physical or social elements to be of unequal relevance to the environmental image, equal numbers of each were used.

A major consideration in designing the grid was the number of constructs and elements to include. The number of dimensions used to describe a set of elements is known to vary with the respondent's familiarity with the elements concerned.[8] Since the house and neighbourhood are very familiar to most respondents, a comparatively large number of constructs seemed appropriate. Although Smith and Leach[9] suggested fifteen to be about the optimum number, Green and Rao[10] reported that use of more than eight scales made the grid subject to decreasing returns of information. As for elements, although some studies within geography have used over twenty,[11] it has been noted that increasing the number of elements from ten to fifteen may markedly reduce the amount of structure which the subject could impose on his or her image.[12]

After two pilot surveys the final version of the RG comprised twelve elements and seven constructs (see Table 8.1). Four versions of the RG were used, varying the element and construct sequences and also the construct polarities to counteract possible bias induced by layout.

Research context

Residential mobility is a suitable research topic through which to study the link between environmental image and decision-making because: (a) the familiarity of people with their own residential environment should produce strong images, and (b) the mobility decision, being non-trivial, involves conscious deliberations, so it should be possible to detect clear trends in decision-making. The field-work was carried out in the Belfast Urban Area (BUA), which houses over 600,000 people. Although 'the troubles' do play a major role in determining to where a mover will move, they rarely figure in the motivations behind the decision to move. Indeed, on the basis of extensive research, it appears that the

Table 8.1 Repertory Grid constructs and elements

Constructs

Satisfactory–Unsatisfactory
Middle-Class–Working-Class
Easy to Cope with–Difficult to Cope with
Cheerful–Drab
Appealing–Unappealing
Modern–Old-fashioned
Close to your Ideal–Far from your Ideal

Elements

House Size
Garden/Yard
House Layout
Structural Soundness of House
Amenities in House
House Running Costs
Outlook of House
Privacy from Neighbours
Immediate Neighbourhood
Neighbours
Noise in Neighbourhood
Main Shopping Facilities

---

residential decision-making processes of Belfast householders are predominantly the same as those of householders anywhere else.[13]

The social survey was to involve only owner-occupiers, because their freedom from residential mobility constraints, and hence freedom in residential decision-making, is considerably greater than for those in other housing sectors. Accordingly, areas of high owner-occupier rates were identified. Inspection of socio-economic status and age of head of household, each variable classified as high, medium, or low, showed that 70 per cent of owner-occupiers were in the three medium socio-economic groups or the middle-age, high socio-economic group. Four areas were therefore required, one representing each major type. High levels of physical and social homogeneity were essential to ensure that each person within a given study area would be responding to similar stimuli. Large-scale maps of potential survey areas were closely inspected with regard to such factors as general regularity of house size and presence of unusual environmental features. Also, only women were to be interviewed. This had the practical advantage that they were more likely than men to be at home during the day. The areas finally selected for cluster sampling were compact (maximum size 500 metres square) and will be referred to as Shankill, Castlereagh, Newtownbreda, and Malone (see Figure 8.2). Figures 8.3, 8.4, 8.5, and 8.6 show typical housing in each area. In all, 203 residents were interviewed.

The study areas displayed little overlap in terms of social-class structure. From lowest to highest the rank order was Shankill, Castlereagh, Newtownbreda, and Malone. This rank order repeated time and again as other objective aspects were considered: education levels, employment status, income levels, car ownership, and so on. The average net annual valuation figures revealed this pattern very clearly: £44, £173, £192, and £307. While the average age of household head was about the same (46) in Shankill

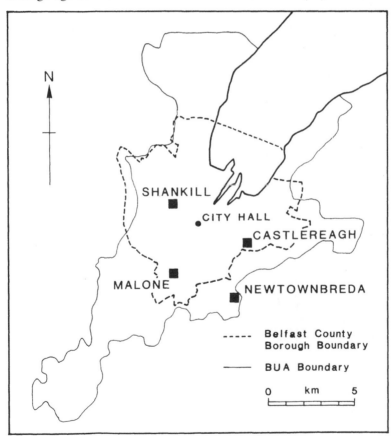

Figure 8.2 Location of study areas

Figure 8.3 Typical Shankill housing

Figure 8.4 Typical Castlereagh housing

140

Figure 8.5   Typical Newtownbreda housing

Figure 8.6   Typical Malone housing

and Newtownbreda, the range in Shankill was substantially larger due to the disproportionately large numbers of young and old household heads, compared with the predominance of the young to middle age category in Newtownbreda. Castlereagh and Malone had rather older household heads (53).

Out of the more than forty subjective variables that were measured, covering a wide range of house and neighbourhood related factors, from convenience to schools to level of personal involvement in the community, and from general tidiness of the neighbourhood to air pollution, the general pattern of responses was highly repetitive, confirming that four substantially different study areas had indeed been chosen. A summarizing procedure has been applied to the original data so that the following figures are the average scores for all these variables set on a scale of 0 to 100. A value of 100 would indicate that every scale had the strongest positive response, and a score of 0 the most negative response. The summary values for Shankill, Castlereagh, Newtownbreda, and Malone are, respectively, 56:81:77:89. This conforms well to the relative levels of environmental quality as judged by the measures above, but the slight superiority of Newtownbreda over Castlereagh is reversed in these perceptual variables due mainly to Castlereagh being an older housing area with a real atmosphere of friendliness.

Within each study area it would have been possible to identify any number of subgroups on which to base consensus grids. On the grounds that the desire to move logically precedes decision and action, but is not subject to all the unforeseen circumstances and constraints which may unexpectedly cause or prevent a house move, two consensus grids were computed for each area: one for those who indicated an intention to stay in their house for at least the next year – the 'stayers', the second for those who indicated an intention to move within the next year – the 'movers'.

Hypothesized aspects of the images

It has already been stated that we should be able to predict at least some of the main features of a typical image using established theories of residential decision-making. By comparing such hypothesized images with observed images, the relevance, or otherwise, of the various concepts and theories should become apparent.

Perhaps the most obvious expectation of the images is that they should accord reasonably closely with the observed conditions. Brown and Moore[14] connect environmental stress (or low residential satisfaction) with the desire to move. If this is true then we should expect movers' images to be less favourable than those of stayers. This will be straightforward to determine by examining construct means, particularly the five which are most clearly of an evaluative nature.

Rossi's work[15] on the role of housing needs in prompting residential moves leads us to anticipate that movers will tend to be more concerned about, and less satisfied with, house size than stayers. To support this contention the measure of relative importance, the 'salience', of the element House Size should therefore be higher among movers, while their general evaluation of this element should be poorer.

It has often been asserted, and introspection would tend to confirm it, that aspirations play an important part in the residential decision-making process.[16] Since aspirations are highly variable and rarely characterized in iron-clad detail, there would seem to be no prospect of ever devising a common metric through which they could be quantified. Nevertheless, because most people have a sense of their ideal they can compare the existing situation with what Pred called their 'dream goal'.[17] If movers are conscious of moving to fulfil an ambition (or at least to get closer to it), then the construct Close to Ideal–Far from Ideal should be more salient among movers than among stayers.

Construct structure of the images

The four pairs of consensus grids were analysed by principal-components analysis and are presented in Figures 8.7, 8.8, 8.9, and 8.10 in diagrammatic fashion. To people familiar with the areas, they provide almost instantly recognizable pictures of each. The main component forms the x axis, and explains between 51 per cent and 78 per cent of the variance. This is very clearly an evaluative dimension in all images, with the five constructs (which, for brevity, will be referred to only by their positive pole for the rest of this chapter), Satisfactory, Easy to Cope With, Cheerful, Appealing, and Close to Ideal, invariably highly loaded upon it. The other two constructs, Middle-class and Modern, generally load heavily on component two, the y axis. As a generalization over all the grids, the two components respectively represent evaluative and status dimensions. Although the exact definition of these components varies across grids, it is evident in the different relationships of constructs to one another (the image diagrams show how the constructs cluster together differently for each image) that there is a strong similarity of components across all the images. As the diagrams indicate, the role of individual constructs

POSITIVE POLES OF CONSTRUCTS

1 Satisfactory
2 Middle Class
3 Easy to cope with
4 Cheerful
5 Appealing
6 Modern
7 Close to your ideal

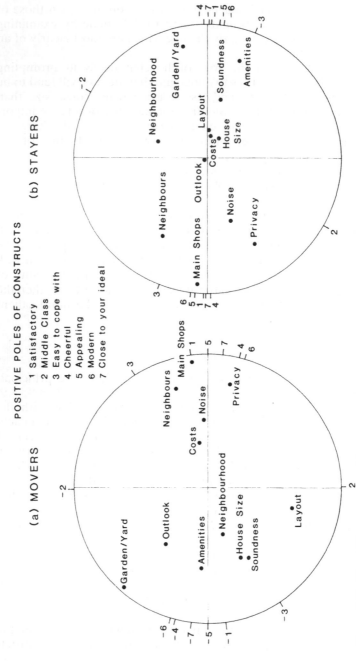

(a) MOVERS (b) STAYERS

Figure 8.7 Shankill consensus grids

POSITIVE POLES OF CONSTRUCTS

1 Satisfactory
2 Middle Class
3 Easy to cope with
4 Cheerful
5 Appealing
6 Modern
7 Close to your ideal

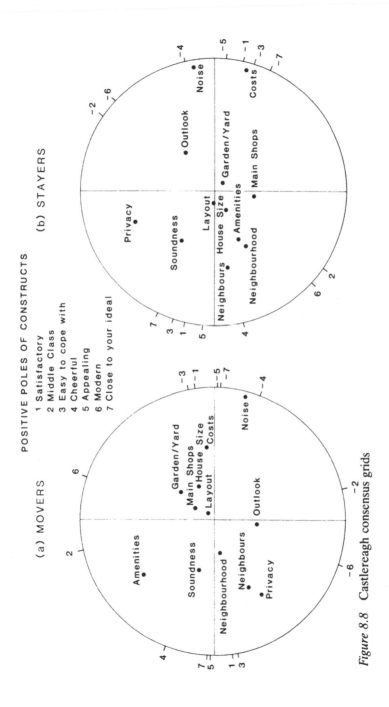

Figure 8.8 Castlereagh consensus grids

POSITIVE POLES OF CONSTRUCTS

1 Satisfactory
2 Middle Class
3 Easy to cope with
4 Cheerful
5 Appealing
6 Modern
7 Close to your ideal

(a) MOVERS

(b) STAYERS

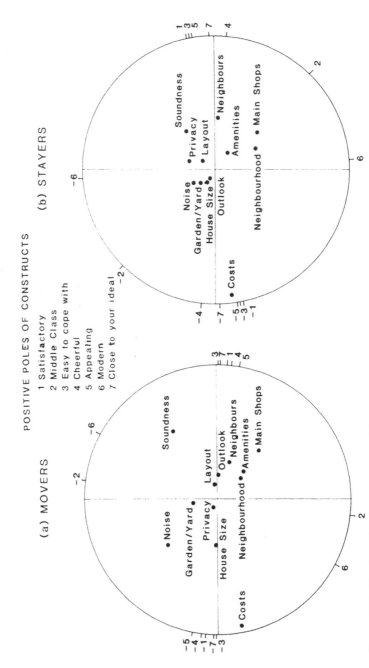

Figure 8.9 Newtownbreda consensus grids

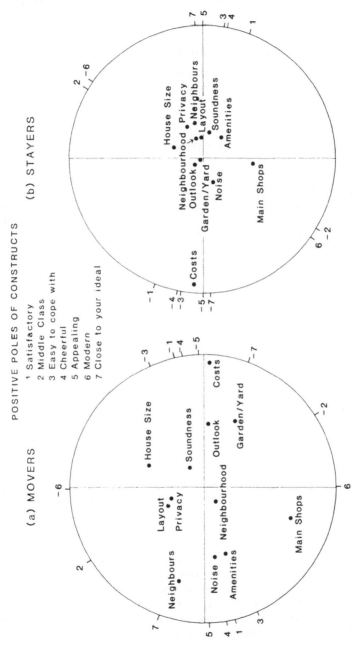

POSITIVE POLES OF CONSTRUCTS

1 Satisfactory
2 Middle Class
3 Easy to cope with
4 Cheerful
5 Appealing
6 Modern
7 Close to your ideal

(a) MOVERS

(b) STAYERS

Figure 8.10 Malone consensus grids

is somewhat more variable, while the role of the individual element is, of course, highly situation-dependent.

Overall levels of image favourability

A comparison of construct means for movers and stayers will locate any differences between their levels of image favourability and so indicate if stress is an important factor in house moving. Using the means for the five constructs which load heavily on the evaluative component (see Table 8.2), the construct means for movers are noticeably higher than for stayers in all study areas, confirming that movers have images which are less favourable overall. This provides clear support for the notion of stress–strain in the residential context and also for the simple idea that a low level of residential satisfaction is associated with the desire to move house. Having concluded that the images of movers are generally less favourable than those of stayers, the inevitable questions arise: first, by how much, and, second, why? Why should there be any disparity in the first place, and why should the size of the disparity vary from area to area?

The rank order of the study areas' desirability is apparent from the construct means, and corresponds with the objective and subjective assessments presented earlier (see Table 8.2). Considering the evaluative constructs collectively,[18] the average difference of construct means between all movers and stayers is 0.522. Why these differences range so much in size, from 0.869 to 0.201, is not obvious.

At the research-design stage one of the major concerns was to ensure a high level of homogeneity of housing conditions within each study area, and the areas were small enough to ensure quite similar neighbourhood conditions for all households in each location. It was expected, therefore, that the size of the differences between the construct means of movers and stayers would have been very similar in all areas.

Given that the neighbourhood-related elements will be virtually the same for everyone within a particular study area, the differences between movers' and stayers' scores might be expected to be large on house-related elements (since individual houses are bound to vary) but comparatively small on neighbourhood-related elements. Plotting element scores on the construct Satisfactory, which loads heavily on component one, and is therefore highly representative of the other evaluative constructs, we see that for each pair of grids there is a very close correspondence of the response patterns which rise and fall almost in step, albeit at different levels of

Table 8.2 Construct means

	Shankill		Castlereagh		Malone		Newtownbreda	
	Movers	Stayers	Movers	Stayers	Movers	Stayers	Movers	Stayers
Satisfactory	4.439	3.045	2.617	1.812	1.800	1.540	2.476	2.088
Middle-Class	5.553	6.053	3.592	2.827	2.417	2.598	2.667	2.867
Easy to Cope	3.348	2.970	2.508	1.908	1.517	1.702	2.220	2.007
Cheerful	4.674	3.726	3.192	2.622	2.167	1.963	2.530	2.224
Appealing	4.508	3.720	2.808	2.124	2.217	1.982	2.649	2.331
Modern	4.773	4.923	3.242	2.868	3.033	3.388	2.089	2.379
Close to Ideal	4.947	4.107	3.158	2.209	2.417	1.930	2.827	2.564
Mean of evaluative constructs (see text)	4.383	3.514	2.857	2.135	2.024	1.823	2.540	2.243
Difference between pairs	0.869		0.722		0.201		0.297	

satisfaction (see Figure 8.11). It is evident that the scores of movers are fairly consistently higher (and therefore less satisfactory) than the scores of stayers on virtually all elements. Patently, it is *not* the case that the house-related elements account for most of the difference between the overall construct means for movers and stayers. Rather, it is as if the respondents have arrived at a general level of contentment with their residential situation which then colours their evaluation of all the elements to a more or less similar extent.

In answer to the question why there should be any disparity in favourability levels, a well-established psychological theory is of help. Cognitive Dissonance Theory,[19] which states that people find it difficult to maintain mutually exclusive attitudes and will strive towards inner 'consonance', helps to explain the markedly consistent nature of the gap between the evaluations made by the two groups within each area. Hence, a resident may find just one or two aspects of her current situation so unsatisfactory as to make her consider moving, but if many other aspects of her residential setting continue satisfactory, she will tend to seek out information which will restore consonance to her attitude set. She will, to reverse the song line, 'accentuate the negative, eliminate the positive'.

Why should the disparity between movers' and stayers' image favourability vary substantially among areas? If we take it as axiomatic that one of the most fundamental ideas that we cherish about ourselves is that we are rational beings, then, although we may tend to downrate our surroundings if we are in a frame of mind to move house (and, vice versa, to rate them quite highly if we intend to stay), there must be some apparent justification for the degree to which we downrate them – there must be some 'negative' to accentuate. The Shankill is an area in poor condition and replete with such 'negatives' – if you are looking for aspects to criticize there is no shortage of them. On the other hand, the Malone neighbourhood is still among the most pleasant and prestigious in Northern Ireland, with no very evident problems. It is not surprising, then, that the evaluative scores here for neighbourhood elements are almost identical for movers and stayers. The movers do downrate aspects of their houses, though not by a great amount, presumably because of the general absence of any serious housing defects. Residential conditions in Newtownbreda are perfectly adequate. If undistinguished and monotonous, most houses are of relatively recent construction, and the standard of house repair is high. Thus, there is limited scope for criticizing the area, and the movers have only a slightly less favourable view of it

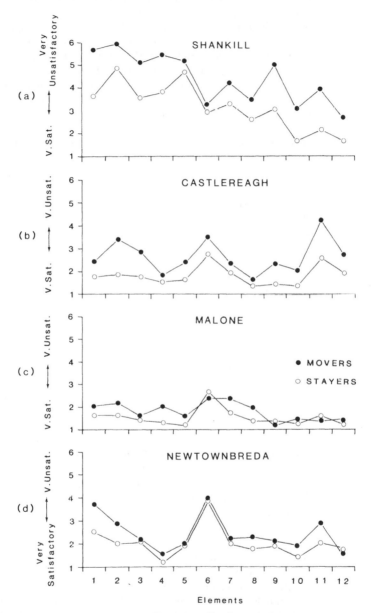

Figure 8.11 Element means for movers and stayers on construct satisfactory–unsatisfactory

than the stayers. By contrast, in Castlereagh the discrepancy between movers' and stayers' views is very evident. This area has housing which is beginning to look rather old and just slightly down-at-heel. The internal house layout, for instance, is old-fashioned, with smallish, somewhat dark rooms, whilst exterior brick surfaces show the effects of 50-odd years of weathering. Most residents have kept their houses in good repair and obviously take pride in their gardens. Unfortunately, a small but noticeable number of properties have received little or no attention in recent years and have become eyesores to varying degrees. There is also a particular local problem of traffic noise on a major road. These and other aspects provide ample justification for those who are looking (consciously or unconsciously) for reasons to lower their evaluation of the area.

To generalize, the degree of difference between movers' and stayers' evaluations of their residential environment is inversely related to the overall quality of the area: the lower the quality of an area, the larger the discrepancy on evaluative scales between would-be movers and stayers. People's environmental images, then, comprise assessments of current conditions viewed with an eye to future intentions.

Element salience patterns

It has been hypothesized that the element House Size should be more important to movers than stayers, and that they will tend to give it a lower evaluation. Taking the second point first, House Size is, overall, the second most unsatisfactory element (after House Costs), and movers have a substantially less favourable view of it than stayers.

The relative importance of an element, referred to as salience in the context of RG analysis, is indicated by its sum of squares as per cent of total variation. The size of an individual value which would obtain if all twelve elements were of exactly equal salience is simply $100/12 = 8.3$ per cent; sums of squares substantially greater than this may be called 'large', and 'small' if much less. A small value implies that the attitude towards an element is indifferent, having been rated near the mean on all the constructs. Conversely, a large sum of squares indicates that a high degree of importance is attached to that element.

The low salience of House Size (see Table 8.3) is quite at variance with what Rossi's work would lead us to expect, and also with the fact that it was so often reported, not only in this study but in a large-scale social survey in Belfast,[20] as the main reason for

Table 8.3 Element salience
(Sum of squares of elements as per cent)

	Shankill		Castlereagh		Malone		Newtownbreda	
	Movers	Stayers	Movers	Stayers	Movers	Stayers	Movers	Stayers
House Size	6.36	1.24	4.39	1.21	5.44	3.68	7.58	1.17
Garden/Yard	15.82	19.00	5.32	1.22	8.72	1.84	2.43	2.12
House Layout	7.46	1.23	2.19	0.39	2.50	3.10	1.60	1.28
Soundness	8.22	5.54	4.44	9.41	4.16	2.59	14.42	8.38
Amenities	6.44	14.86	13.70	4.49	7.10	4.76	3.84	2.12
House Costs	4.69	2.36	10.48	23.08	22.21	55.14	37.24	50.98
Outlook	4.70	0.14	4.88	6.78	10.92	0.76	1.33	0.82
Privacy	10.51	13.19	12.55	10.67	3.95	5.06	2.20	2.68
Neighbourhood	3.82	3.77	2.73	5.86	3.31	1.88	5.20	7.47
Neighbours	11.05	10.77	9.92	9.39	13.10	6.01	3.44	9.87
Noise in Neighbourhood	4.77	6.56	27.79	24.76	7.60	5.96	11.27	2.55
Main Shops	16.16	21.34	1.60	2.74	10.53	9.22	9.46	10.56

moving house. This may be explained by recognizing that the terms importance and salience are not quite identical. In psychology, salience refers to the thoughts about a certain subject which come most readily to mind.[21] In considering the location of a new house, for example, most people would set some store on having reasonable access to hospital facilities. However, the mother whose child had to be admitted to a casualty department last week, if asked what she considers important about a new house, will almost certainly place proximity to a hospital very high on her list of priorities. It is suggested, tentatively, that the salience patterns may in part be accounted for by the amount of thought people are likely to have given to certain elements in the last few days. The elements with the lowest overall salience are surely not unimportant to residents; they include House Size, House Layout, Outlook, and Neighbourhood. To change such aspects, however, would require either significant house alterations or a house move. The time-scale involved is fairly long before action could be taken, and for only a few people would these represent problems which had become in any way acute.

These salience patterns suggest that the environmental image, not surprisingly, strongly reflects current circumstances. Nevertheless, that the images are also influenced by people's intentions is evidenced by the different favourability levels noted earlier between movers and stayers. This accords also with the salience values for House (running) Costs which broadly match actual running costs: low for small houses, high for large ones. Excluding Shankill, where there are such limited outgoings on houses, House Costs is much more salient and negatively rated among stayers than among movers – by a factor of over 2 in Castlereagh and Malone, and of about 1.4 in Newtownbreda. From this we may conclude that movers tend to be much less concerned about the expense of running their house and, by implication, about financial constraints in general and the difficulties of moving house in particular. It seems, therefore, that respondents who expressed a desire to move were also those who saw a real prospect of being able to do so. This provides clear evidence for the widely held view that aspirations tend to conform to the level of the attainable.[22]

The role of aspirations

If aspirations are more to the fore in a resident's mind when a house move is being considered, then the construct Close to Ideal should be more salient among movers than stayers. This is based on the assumption that the 'ideal' is reasonably synonymous with the level of aspiration. The relative importance (or salience) of constructs is gauged by their variation as per cent of the entire variation in a grid. With seven constructs, a completely random response set would yield $100/7 = 14.3$ per cent variation per construct, so this figure is the standard against which a construct's relative importance is assessed.

Table 8.4 shows that the average salience values of Close to Ideal for all movers and stayers are high at 20.50 per cent and 17.06 per cent respectively. (Values in each area differ little from these overall figures.) The difference between the two values is the largest for any of the constructs, showing that Close to Ideal is especially important to movers, and suggesting that it may indeed be of value in detecting aspirations.

Table 8.4 Construct salience
(Variation of constructs as per cent)

	Movers	*Stayers*
Satisfactory	13.02	13.07
Middle-Class	3.15	6.22
Easy to Cope	11.64	11.84
Cheerful	16.92	16.28
Appealing	21.79	23.28
Modern	12.98	12.75
Close to Ideal	20.50	17.06

Aspirations are very difficult to measure in any precise way as they are often nebulous, ill-formed, and idiosyncratic. From the previous section, it is apparent that those elements which one might expect to be closely related to long-term aspirations, such as House Size or Neighbourhood, are rarely salient. Therefore an extension to the RG analysis was employed in order to detect aspirations.

Accepting that aspirations tend to conform to the level of the attainable, then those conditions which seem attainable become the standards against which movers will appraise their current circumstances. On this basis it is suggested that elements which movers downgrade may be equated with their aspirations, though the question arises of how much downgrading indicates an aspiration. As there is no single element score against which such downgrading can be assessed (given that element scores vary with environmental conditions), the up- or downgrading in movers' scores will be determined by taking the stayers' scores in each study area as reference points.

The method of grid comparison to be used is implemented by

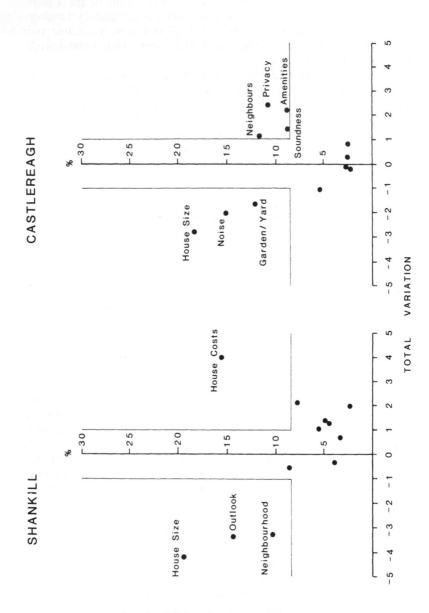

SUMS OF SQUARES AS PERCENT

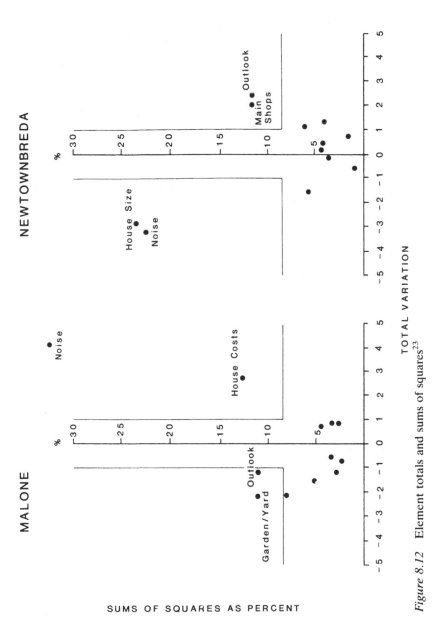

NEWTOWNBREDA

MALONE

TOTAL VARIATION

SUMS OF SQUARES AS PERCENT

Figure 8.12 Element totals and sums of squares[23]

DELTA.[24] This computer program is designed to compare two grids aligned by construct and element, and is therefore ideally suited for use with a pair of mover/stayer grids. DELTA forms a single new grid by first rewriting the original grids in terms of deviations from construct means, and then subtracting the first grid (movers) from the second (stayers). The new grid, called the DELTA grid because it is composed of the changes from one grid to the other, is then analysed by an abbreviated form of the INGRID procedure. Interpretation of the results revolves around two values. The first value is total variation, where a large figure indicates that an element tends to be scored consistently differently on the two original grids in terms of all the constructs. The sign of the total variation figure reveals whether it is usually upgraded (+ve) or downgraded (−ve) on the first grid compared with the second. Thus, a negative total variation indicates that movers tend to downgrade an element more than stayers, and vice versa. The second value is sum of squares as per cent, where elements with the highest values differ most between the grids. As there are two values involved, it is necessary to identify some thresholds above which we may begin to feel confident that the result is due to more than random error. Given that a sum of squares value of 8.3 per cent would obtain for each element under completely random conditions, this is the obvious choice as a threshold value. Unfortunately, there is no comparable theoretical basis for identifying a similar threshold value for the total variation figures. The threshold value of +1.0 was chosen after inspection of relevant measures of central tendency and dispersion in conjunction with visual inspection of the graphed data. Only elements with values over both these thresholds are labelled on the diagrams (see Figure 8.12).

It is evident that movers consistently downgraded House Size, often to a considerable extent. Only in Malone is this phenomenon attenuated, almost certainly reflecting the already generous proportions of most homes there. There seems good reason in this case to accept that the downgrading of House Size does point to movers' aspirations.

In both Castlereagh and Malone, Garden/Yard is consistently downgraded by movers, implying the existence of at least a modest aspiration. Additional information from a questionnaire confirms that all movers in Malone want a smaller garden. The absence of downgrading on Garden/Yard implies that it does not represent an aspiration in Newtownbreda or Shankill. Since garden and yard space is much the same in Newtownbreda and Castlereagh, it is not clear why Garden/Yard should emerge as an aspiration in one area and not in the other. That Garden/Yard appears not to equate to an aspiration in Shankill is rather surprising since none of the houses there have gardens, just tiny and highly unsatisfactory enclosed yards.

There is a strong and very consistent downgrading of Outlook by Shankill movers – definitely an aspect they wish to improve. In Malone, Outlook is somewhat downgraded by movers, perhaps because it is one of the few aspects of housing in that area that could be better.

If the goal of status enhancement is an important aspiration for movers,[25] then we would expect movers strongly to downgrade Neighbourhood, but this occurs only in Shankill. This, together with the fact that the construct relating to class, namely Middle-Class, is overall the weakest of the constructs for both groups (Table 8.4), points to the conclusion that the desire for increased social status is, at best, a second-rank motivation for moving house.

That there is a particular problem in Castlereagh with Noise in Neighbourhood has been noted earlier, and it is clear from the DELTA results that movers in this area are both consistent and emphatic in their desires to live in a quieter locality. The same is also true for Newtownbreda where movers are quite noticeably less satisfied with the noise level than stayers.

All of the aspects identified above as aspirations correspond to aspirations strongly anticipated by the author on the basis of quite lengthy field-work. Consistent interpretation of element downrating, however, would imply the existence in Newtownbreda of an unanticipated aspiration, namely Noise in Neighbourhood, which, from a knowledge of the area, is difficult to justify. It appears, therefore, that examination of element downgrading provides a good, though not infallible, indication of group aspirations.

Conclusion

In this chapter an attempt has been made to measure a limited portion of the behavioural environment relevant to residential decision-making. The measurement technique, based on Personal Construct Theory, produced consensus grids for groups of would-be movers and stayers in four substantially different owner-occupier areas in Belfast. Since grids are primarily descriptive tools and interest here lies mainly in decision-making processes, existing ideas and theories about residential decision-making were used to hypothesize facets of image content and structure. Where images conformed to these expectations it was taken as evidence that particular decision-making processes were at work. This approach also led to valuable observations about the nature of environmental images.

The influence of stress in encouraging mobility found confirmation in the different levels of image favourability displayed by movers and stayers. Patterning of the evaluative scores for individual elements was striking: nearly all the elements in any grid seem to be influenced by a general evaluative level. The consistent difference in evaluative scores between movers and stayers in all four areas strongly indicates that this result is no sampling fluke, but a widespread phenomenon.

Initially it was assumed that the parts of the image particularly relevant to decision-making would correspond to the most salient elements. On the contrary, it emerged that higher salience levels tend to be associated with elements which represent the immediate problems and difficulties of day-to-day living.

It took the sophistication of DELTA grids, which compared movers' and stayers' images, to detect elements which are involved in longer-term thinking. While people form their image of the current environment with their aspirations in mind, the image is predominantly present- rather than future-orientated.

Although intention to move house has been used in this study as a surrogate for actual mobility, the relevance of the images to future spatial behaviour was established 12 months after the survey on which this chapter is based. Fully 24 per cent of the would-be movers had moved, and a further 27 per cent indicated that they had come close to moving. Figures for would-be stayers were just 5 per cent and 4 per cent respectively.

It has become apparent that we have in the Repertory Grid technique a powerful methodology for measuring the perceived world. By combining this technique with Kirk's insight into the importance of understanding decision-making processes, implicit in the behavioural environment concept, it has been possible to detect, sometimes clearly, sometimes faintly, some of the processes that ultimately transform perception into action.

Notes

1 W. Kirk, 'Problems of geography', *Geography*, vol. 48 (1963), pp. 357–71.
2 Ibid., p. 370.
3 Ibid., p. 366.
4 T. J. Anderson, 'Environmental perception and residential decision-making', unpublished Ph.D. thesis, Department of Geography, Queen's University, Belfast, 1985.

5 G. A. Kelly, *The Psychology of Personal Constructs*, vols 1 and 2 (Norton, New York, 1955).

6 P. Slater, *Notes on INGRID 72* (Academic Department of Psychiatry, St George's Hospital, London, 1972). For detailed explanation on diagrammatic presentation of INGRID results see P. Slater, *Composite Diagrams and Systems of Angular Relationships Applying to Grids* (Academic Department of Psychiatry, St George's Hospital, London, 1973).

7 The theoretical basis is discussed in P. Slater (ed.), *The Measurement of Intrapersonal Space by Grid Technique. Vol. 2: Dimensions of Intrapersonal Space* (John Wiley, London, 1977). A practical guide to the computer analysis is in P. Slater, *The Grid Analysis Package of Computer Programs* (Academic Department of Psychiatry, St George's Hospital, London, 1977).

8 D. Bannister and J. M. N. Mair, *The Evaluation of Personal Constructs* (Academic Press, London, 1968).

9 S. Smith and C. Leach, 'A hierarchical measure of cognitive complexity', *British Journal of Psychology*, vol. 63 (1972), pp. 561–8.

10 P. E. Green and V. R. Rao, 'Rating scales and information recovery – how many scales and response categories to use', *Journal of Marketing*, vol. 34 (1970), pp. 33–9.

11 J. A. Harrison and P. Sarre, 'Personal construct theory in the measurement of environmental image', *Environment and Behavior*, vol. 7 (1975), pp. 3–58.

12 Bannister and Mair, *The Evaluation of Personal Constructs*.

13 F. W. Boal and M. A. Poole, 'Religious residential segregation and residential decision making in the Belfast Urban Area', Final Report to Social Science Research Council, on Research Grant 1165 1/2, 1976.

14 L. A. Brown and E. G. Moore, 'The intra-urban migration process: a perspective', *Geografiska Annaler*, Series B, vol. 52 (1970), pp. 1–13.

15 P. H. Rossi, *Why Families Move: A Study in the Social Psychology of Urban Residential Mobility* (Free Press, Glencoe, Ill., 1955).

16 A very influential example is A. Buttimer, 'Social space and the planning of residential areas', *Environment and Behavior*, vol. 4 (1972), pp. 279–318.

17 A. Pred, 'Behaviour and location: foundations for a geographic and dynamic location theory', part 1, Lund Studies in Geography, Series B: Human Geography, no. 27 (1967), Gleerup, Lund. See also W. Michelson, *Environmental Choice, Human Behavior and Residential Satisfaction* (Oxford University Press, New York, 1977).

18 Only the five evaluative constructs, for which a low value indicates a more favourable evaluation, are taken into consideration. For the other two constructs, Middle-Class and Modern, there is probably only incomplete agreement about which end of each scale is intrinsically more favourable.

19 For a lucid explanation and critical appraisal of Festinger's theory see B. Reich and C. Adcock, *Values, Attitudes and Behaviour Change* (Methuen, London, 1976).

20 Boal and Poole, 'Religious residential segregation and residential decision making'.
21 M. Tuck, *How Do We Choose?* (Methuen, London, 1976).
22 G. L. S. Shackle, *Decision, Order and Time in Human Affairs* (Cambridge University Press, Cambridge, 1969).
23 See Note 18.
24 Slater, *Dimensions of Intrapersonal Space*, pp. 149–53.
25 W. Bell, 'Social choice, life styles, and suburban residence', in W. F. Dobriner (ed.), *The Suburban Community* (Putnam's, New York, 1958), pp. 225–47.

Chapter nine

Outrage and righteous indignation: ideology and imagery of suburbia

John R. Gold and Margaret M. Gold

By tradition, the suburb has enjoyed a poor reputation among urban commentators and practitioners in Great Britain and North America; indeed, the examples of derogatory quotes that can be assembled is literally endless. We can read of the spread of suburbia being likened to the spread of the 'brown rat, ravaging everything before it'.[1] The semi-detached British suburban house has been described as 'perhaps the least satisfactory building unit in the world',[2] with its design being castigated as the 'neo-Nothing',[3] bought for the 'social cachet that . . . [it is] presumed to bring'[4] and its 'repetition' being held to produce an 'inescapable monotony of mass'.[5] The epiphet 'mass' is particularly popular in writings about suburbia, for example, with Robert Sinclair[6] arguing that associated 'with 'the mass-produced home is the mass-produced man'. Suburban estates, then, are argued to be a failure 'from a sociological as well as from an aesthetic point of view'.[7] They are held to be 'lifeless, culturally barren',[8] 'little textbooks of social sterility'.[9] 'Suburbia', wrote Lewis Mumford, 'offers poor facilities for meeting, conversation, collective debate, and common action – it favours silent conformity'.[10] More worrying still, perhaps, we can read of the dangers of the future city as becoming a glorified suburbia occupied by 'robot-like human beings, differentiated only by their identification numbers, who live in Levittowns where homes are differentiated only by their street numbers'.[11]

Without much difficulty, it is possible to assemble an imposing body of literature that holds that suburbia is a visually impoverished environment that is also socially and morally impoverished. It is a literature which possesses a remarkable sense of detachment: you would hardly believe that what is being described is one of *the* typical landscapes of the contemporary city, the living environment for around half the British population,[12] and an environment which studies of housing preferences have consistently shown to be

163

highly regarded by the general public.[13] As yet, only a comparatively small number of studies by local historians, social
scientists, and architectural commentators have seriously attempted
to redress the balance.[14] Such works have succeeded in analysing
why urbanists have routinely written in this manner and have
made attempts to look beyond élite prejudice against suburbia in
order to build a more accurate picture of the subject. Yet in terms
of creating new directions for enquiry, they have, with one or two
notable exceptions, done little more than to provide an outline
agenda for further research and to portray the possible counter-
imageries of the diversity of suburbia and the richness of suburban
life. By comparison, there have been few attempts to examine in
any detail the nature or function of the imagery that has been put
forward.

With this in mind, this essay adopts a different approach to most
previous work in this field. Conceived as an exploration of what
Kirk[15] termed the behavioural environment, it demonstrates the
way in which members of the architectural profession and professional architectural commentators built up, consciously and
otherwise, an adverse imagery of suburbia, and examines the
continuing implications of that imagery for the conduct of urban
studies and for our interpretations of the contemporary city. The
method chosen is to begin with a case-study as a means of
understanding the origins and anatomy of anti-suburban imagery.
The example selected is *Outrage*,[16] an influential, and stridently
anti-suburban, monograph by the late English architectural critic
Ian Nairn which first appeared in the mid-1950s. After briefly
reviewing its contents, the essay will place *Outrage*'s anti-suburban
imagery in a wider media context, focusing on its use of architectural photography. The ensuing section of the chapter casts the
discussion still more widely by relating this imagery to its ideological context, before the final part draws together the major threads
of the discussion.

Outrage

Outrage was first published in June 1955 as a special issue of the
monthly journal, *Architectural Review*, and was re-issued the
following year by the Architectural Press as a monograph in its
own right. *Outrage* was to be followed in 1956 by a sequel, a
collection edited by Nairn entitled *Counter-Attack against Subtopia*,[17]
and was the precursor of a long-lived column that Ian Nairn came
to write for the *Review*.

Outrage took its name from a statement by Sir George Stapledon:

Figure 9.1 'Agents of Subtopia'

Source: I. Nairn, 'Outrage', *Architectural Review* (special issue), vol. 117 (1955), p. 370. Reproduced with the permission of the *Architectural Review*.

'It is an outrage on posterity to misuse a single yard of land – the outrage has been more than sufficiently perpetrated already.'[18] Nairn was concerned with a wide variety of practices that were leading to the aesthetic degradation of the environment. His targets were groups involved in the process of urban development –

speculative builders, advertisers, public utilities, and local bureaucrats – but, significantly, architects did not feature in his criticisms.
Aided and abetted by his illustrator Gordon Cullen (see Figure
9.1), Nairn mounted a fierce attack on such matters as the
proliferation of pylons and overhead wires, ill-placed advertising
hoardings, poor design of street fitments, land-hungry road lay-
outs, and, in particular, the chaotic scenery on the growing
fringes of towns caused by the absence of adequate control over
development.

After a brief prologue in which terms are outlined and key
arguments posed, the heart of the study comprises ninety pages of
photo-journalism. It consists of a travelogue of a journey through
the kingdom, and, as such, follows in the tradition of Celia
Fiennes, Daniel Defoe, and J. B. Priestley. Nairn provided an
illustrated report of a road journey from south to north, or, to be
precise, from Southampton to the Scottish Highlands, cavalierly
omitting the Southern Uplands and Central Lowlands of Scotland
– apparently for reasons of space. The message that comes across
is a powerful one. Forces of development were running rampant:
they were eroding the visual quality of the environment, they were
eroding the distinctiveness of place, they were fudging the distinctiveness of town and country and inflicting formless chaos in its
place. It was necessary to promote greater awareness and exert
greater control before it was too late. The agencies of environmental degradation that were catalogued included traffic congestion,
rotting aerodromes, pylons, poor or obtrusive signposts, and ugly
forestry developments. Yet, while the attack was ostensibly broad-
based, the commentary and illustrations clearly revealed that
suburbia, described as the 'amorphous destroyer' of Britain,[19] was
the prime target. This is made abundantly clear by the term that
Nairn coined to describe the chaotic landscape of the urban fringe
and beyond – 'Subtopia'.

Subtopia is a composite word derived from 'suburbia' and
'utopia'. It was defined as: 'making an ideal of suburbia. Visually
speaking, the universalisation and idealisation of our town fringes.
Philosophically, the idealisation of the Little Man who lives
there.'[20] There is no mistaking the tone or the sneer implicit in this
statement. The phrase 'the idealisation of the Little Man' is a
telling one, however, because it indicates that what was at
stake was more than just the imagery of the built environment:
there was also a particular imagery of society. As Nairn un-
ambiguously asserted, 'buildings affect people, and Subtopia produces subtopians'.[21]

The exact character of the 'subtopian', or the 'Little Man', was never actually defined, but the following perhaps gives some indication of Nairn's thinking:

> The environment is an extension of the ego, and twentieth-century man is likewise busy metamorphosing himself into a mean – a meany – neither human nor divine. And the thing he is doing to himself and to his background is the measure of his own mediocrity.[22]

The list of epithets that Nairn used is, thus, striking. When considering these and other statements from *Outrage*, one can find suburbia and suburban life described, *inter alia*, as 'amorphous', 'blighting', 'bogus', 'careless', 'dismal', 'homogeneous', 'tedious', 'universal', and 'the product of mediocrity'.

The impact of the text was reinforced by the photographs and illustrations to which the text is attached, a representative sample of which is shown in Figures 9.2–9.4 inclusive. Collectively, they constructed a visual imagery that was wholly hostile to the suburb. The aerial photograph shown in Figure 9.2, entitled 'Subtopia on

Figure 9.2 'Subtopia on the March'

Source: I. Nairn, 'Outrage', *Architectural Review* (special issue), vol. 117 (1955), p. 369. Reproduced with the permission of the *Architectural Review*.

the March', depicts a low density private-sector housing estate with the old field patterns apparently still trying to proclaim themselves. The photograph's caption asserted:

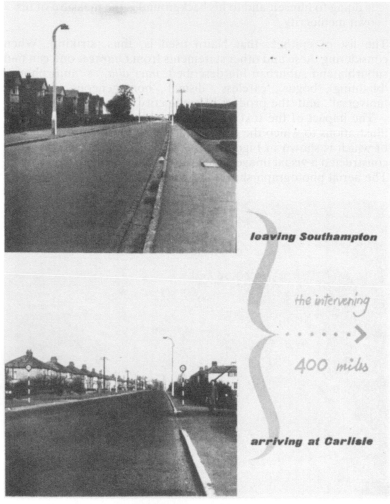

Figure 9.3 'Leaving Southampton ... Arriving at Carlisle'

Source: I. Nairn, 'Outrage', *Architectural Review* (special issue), vol. 117 (1955), p. 391. Reproduced with the permission of the *Architectural Review*.

The whole country is beginning to look like this, as all development follows the same pattern of a careless romp of dispersal across the countryside, a pattern which is wasteful and dreary in itself and utterly impracticable in a small and crowded country.[23]

From Southampton to Carlisle (Figure 9.3), it seemed that the world was becoming ever-increasingly similar, a world of similar, monotonous suburban dwellings and endless tarmacadam, offering landscapes of the type that the suburban estate at Bletchingdon, Oxfordshire (Figure 9.4) could be held to be typical.

Figure 9.4 'Sprawl on the Outskirts', estate at Bletchingdon (Oxfordshire)

Source: I. Nairn, 'Outrage', *Architectural Review* (special issue), vol. 117 (1955), p. 383. Reproduced with the permission of the *Architectural Review*.

Yet the curious, stylized nature of these photographs is readily apparent. If one considers further the composition of the photograph in Figure 9.4, which is of a type used repeatedly in *Outrage*, one can see that it is taken in a particular way. The photographer has selected a day with overcast weather conditions, has stood, or possibly crouched, in the centre of the road and used a wide-angle lens. Thus accomplished, the photograph that we see includes as much flat, boring road surface and featureless sky as possible. The buildings and all human details are pushed to the background.

The question then arises as to the origins of this style of photography and its associated imagery. To answer this requires some insight into the broader context of architectural journalism. It in no way detracts from Nairn's importance as an architectural commentator and protagonist for higher environmental standards to say that *Outrage* did not contain anything particularly new in its imagery of suburbia. The formula already existed. Much the same type of photography and style of writing about suburbia and suburban society could be found in the *Architectural Review*, and in kindred publications, at any time over the previous 20–25 years. It is to this broader context of Nairn's work, and specifically the medium of architectural photography, that we now turn.

Architectural Review

The *Architectural Review* itself will need very little introduction to anyone familiar with architectural journalism and, in particular, with the development of the British Modern Movement. From around 1930 onwards, the *Review* was in practice the mouthpiece for that Movement, a source of support in the face of an architectural press that was either indifferent or hostile to modernism. Part of the *Review*'s significance in the evolution of architectural journalism lay in the rare collection of the talents of the architectural and artistic avant-garde that it succeeded in attracting to its columns. Paul Nash, John Betjeman, Robert Byron, Nikolaus Pevsner, J. M. Richards, Herbert Read, and Hugh Casson all wrote regularly for the journal, with illustrations being provided by artists as notable as John Piper, Osbert Lancaster, Laszlo Moholy-Nagy, and Gordon Cullen. Yet the *Review* was also significant for a series of important advances in the fields of typography, reporting policy, and illustration. For our purposes perhaps the most important lay in its compelling use of photography. During the 1930s, the journal's photographers, particularly M. O. Dell and H. L. Wainwright, effectively turned the architectural photograph into a powerful weapon of propaganda.[24]

To see how this was accomplished, it is instructive to examine an example of what occurred when the photographers of the *Review* wanted to convey a positive view of a building. Figure 9.5 shows a house by F. R. S. Yorke, at High Street, Iver, and is typical of the

Figure 9.5 Photograph by M. O. Dell and H. L. Wainwright of a house at High Street, Iver (architect F. R. S. Yorke)

Source: Architectural Review (1936), p. 80. Reproduced with the permission of the *Architectural Review*.

output of the journal during the 1930s when depicting photographs of architecture which met with editorial approval. Using standard large-format cameras and slow, but high-resolution film (such as HP4, Pan-F), these photographers would use heavy red and

171

orange filters to get the type of effect shown here. It was an effect that flattered buildings and delighted architects. Taken soon after construction, before the rain and drip of the British climate had had an opportunity to streak the façade, the walls were made to seem dazzling white rather than their true colour of dull greyish-white. The contrast is heightened by the deep tones and dramatic cloud-scapes, against which the building stands out ravishingly.[25]

The effect was perfectly deliberate, being intended to lend the building an heroic air. There is absolutely no technical reason why exactly the same strategy could not be used on the average suburban villa or even semi-detached house: after all, the building shown here may meet certain criteria as to materials, geometry, and building style, but for all that it basically remains a suburban villa. The same techniques, however, never were used in any such way. The *Review* reflected the international Modern Movement's deep dislike of suburbia, symptomatized perhaps by Le Corbusier's dictum: '[the suburbs are] often mere aggregations of shacks hardly worth the trouble of maintaining'.[26] In response, the journal adopted entirely different photographic conventions when dealing with suburban topics, centring primarily on two elements, the 'unflattering angle' and the aerial photograph.

The former needs little further discussion since its principal features have already been identified above (Figures 9.3 and 9.4). In this style of photograph, the angle of regard, weather conditions, and materials were painstakingly selected to show the buildings in the least favourable light. Every step had been taken to reduce the ambiguity that is present in all photographs in an effort to anchor its content in the specific meaning that the photographer wished to impart.[27]

Its counterpart, the aerial photograph, effectively performed the same function. When used individually rather than in stereo-scopic pairs, aerial photographs reduce matters to a two-dimensional study of structure and layout. Inevitably, it is the roads, the overall geometry, and the use of land that catch the eye. This type of photograph abounds in ambiguity since the reader is seeing a subject from a totally unfamiliar angle, with few reference points of local detail, relief, or even scale upon which to make judgements or reach conclusions. That ambiguity, however, can be capitalized upon by the caption-writer, since, in the absence of other cues, readers are more prone to accept the characteristics highlighted by the caption-writer, which in this case invariably alluded to low density, wastefulness, regularity, formlessness, monotony, and anonymity.[28]

Taken together, these two forms of photograph comprise the

dominant forms of representation of suburbia to be found in the *Architectural Review*. Viewed analytically, it is clear that there was a distinct visual language being employed; a language that was later reflected and, in some ways, embellished by Nairn's monograph *Outrage*. Yet any interpretation of the imagery portrayed must necessarily confront the question as to why a journal, which was a world leader in architectural photography and produced wonderful, heroic pictures of buildings when it wanted to, should choose to resort to such apparently dismal photographs when dealing with suburbia – either purchased air photographs or seemingly technically and compositionally inferior snapshots. In pursuit of the answer, we must now move to the wider context of 'ideology'.

Imagery and ideology

'Ideology' is a highly contentious term, with a variety of current meanings.[29] At the outset, then, it is important to note that the usage of the term here does not carry the pejorative overtones that can be found, for example, in most Marxian literature on the subject, in which ideology is taken as a distortion of objective reality. Rather, ideology is defined as pervasive sets of beliefs, ideas, and images that groups employ to make the world more intelligible to themselves. Ideology is thus viewed as an essential part of that process by which people come to terms with the world around them and can be regarded as a frame for the experience of individual group-members, leaders, and followers alike. Whether or not that frame is a perversion depends on the content of the ideology rather than the existence of ideology *per se*.[30]

Having said this, it is important to stress two points. First, ideology invariably functions as a conservative force. The effect of an ideology, even amongst an avant-garde and ostensibly radical group such as the architects of the English Modern Movement, is to bring about conformity, relating the thinking of individuals to that of the group as a whole in such a way that individual and group thinking tend to be congruent and mutually reinforcing. Second, it must be realized that an ideology normally serves to support and further a group's own interests. The Modern Movement was caught in the dilemma of wanting to proselytize for new values and awareness,[31] yet urgently needing clients if they were to establish and maintain expensive architectural practices. Although they felt themselves to be outsiders to the established architectural profession, and indeed cultivated the feeling of being outcasts,[32] they effectively had no option but to seek work from those with the funds to offer commissions – invariably members of the established order. Put another way, the Modern Movement had to strive to make its ideas more marketable to that order. Hence, despite rhetoric to the contrary, the Modern Movement's expression of new ideas had the effect of advancing the claims of the modern architect to a greater share of the organization of, and resources available to, the architectural profession.[33]

Turning to the visual imagery of the Modern Movement with these points in mind, it is necessary to begin by recognizing that the *Architectural Review*, of course, did not invent criticism of or hostility to suburbia, and that it, like any other journal, was party to the mood of its time. There was certainly a broad current of intellectual opinion hostile to suburbia, which had already crystallized into coherent form by the early inter-war period, primarily in response to the explosive growth of privately (or speculatively) built suburban housing. Tremendous pressures had been placed on land at the urban fringe, most notably around London, where expansion of suburban housing was fuelled by the upsurge in demand for home ownership amongst the rapidly growing professional middle classes. These pressures themselves were assiduously cultivated by property developers, estate agents, railway companies, and other bodies with a vested interest in encouraging suburban expansion.[34] As a result, for example, the population of certain districts to the north and west of London more than doubled during the inter-war period.

The haphazard pattern of growth that followed undoubtedly provided cause for opposition on planning, conservation, and strategic grounds. Nevertheless, criticism went far beyond these grounds by indulging in the wide-ranging prejudices against the design of suburbia and the character of suburban life that have been identified above. And when one looks through this material, it is an inescapable conclusion that opposition to suburbia had become a standard part of the ideology of a great number of groups. Indeed, as Alan Jackson wryly remarked: 'Among the intelligentsia, *suburb* was already a dirty word [by 1930]. . . . Only the dead and the half-dead were attracted to such places.'[35]

Moreover, what is most impressive is the sheer range of the groups that could find suburbia ideologically repugnant – an unholy alliance of interests, from all parts of the political spectrum, who would surely have been united by precious little else. Quite apart from the Modern Movement, these included rural preservationists concerned about what they saw as threats to the countryside; landlords and estate owners concerned about the threat to property; farming interests; a motley collection of literati, who basically believed that suburban life was a poor substitute for the full urban experience; a variety of protagonists for town planning who abhorred what they saw as uncontrolled sprawl; and the Garden City movement which, besides abhorring uncontrolled sprawl, was desperately uneasy that it would be tarred by its enemies as the advocates of suburbia.

Yet of all these groups the Modern Movement had the most clearly articulated and comprehensive reasons for its hostility to suburbia. In a recent interview recorded by one of the current authors with a former editor of the *Architectural Review*, three such reasons were suggested.[36] The first was connected with the Modern Movement's concept of the town. The Modern Movement strongly favoured the compactness of the town and favoured a sharp divide between town and country. The sprawl of suburbs, the indeterminate merging of town and country was itself regarded as an anathema. The second reason was connected with building styles and materials. The Modern Movement were interested in new materials – like steel, glass, reinforced concrete, sought new building techniques, and wanted a break with the aping of past traditions in building design. Suburbia as the land of brick, timber and tile, traditional building methods, and stylistic allusion was again the very antithesis of what they desired. The third reason was that suburbia was identified with 'bourgeois taste'. To understand that point fully, one needs to have some inkling of the place that the word 'bourgeois' had in the vocabulary of modernism. To be identified as bourgeois was to be establishment, narrow, conservative, and unenterprising. These were the people who would not understand modern design and modern art; they were conventional in their visual and aesthetic ideas. In short, to pick up the term used by Ian Nairn, these were the Little Men. The Modern Movement, by contrast, burned with its own conceptions of modern urban life, its values and organization, and these were different from anything that they could identify in suburban life.[37]

These, then, were the stated reasons for the modernist's deep hostility to suburbia, but it is not difficult to recognize the underlying ideological dimension that was at work here. The Modern Movement of the 1930s felt themselves to be a small and embattled minority struggling against the odds to propagate new forms of building and concepts of city design. The major professional institutions were deemed to be in the hands of traditionalists and plans for modern buildings, even for small and secluded villas, could meet with vehement opposition. The expression of radical new views effectively constituted an assault on the citadel, an attempt to secure a greater hearing for modern architectural ideas

175

and to gain more work and influence for modern architects. Yet such strategy would only succeed if a sufficiently damning attack could be mounted on accepted practices and conventional thought. Considered in such a light, suburbia and suburban life-styles stood four-square with neo-Georgian banks and Queen Anne-style country houses as objects against which righteous indignation could be directed.

Further manifestations of the operation of ideology are apparent from the nature of the Modern Movement's discourse about suburbia. First, it should be stressed that, after the initial years, there was rarely much discussion about the grounds on which suburbia was criticized. To return to the interview mentioned previously,[38] after outlining the broad reasons for the Modern Movement's dislike of suburbia, the interviewee was at pains to point out the following: 'These were the general ideas that were about at that time and we all held them. At the start, I suppose, they were matters for debate, but after that I can't say that we ever spent much time in discussing them.' That in itself is typical of an ideology. An imagery becomes established and becomes part of the reliable background to the group's discourse: once established it is rarely debated.

Second, it should also be emphasized that ideology does accumulate incrementally over time. Another respondent, himself a key figure in the London County Council's Reconstruction Areas programme after the Second World War, suggested that differences in thinking between the old and new generations of modern architects did progressively emerge. Pointing to the importance of the divide between pre- and post-war modern architects, he noted the following: 'The things that we [i.e. those trained before the war] argued about, they [i.e. those trained after the war] took for granted. It was the basis on which they began.'[39]

Discussion

We have come a long way from our starting position of a discussion of Nairn's monograph *Outrage*, but have now acquired the basis from which one can interpret the imagery of suburbia that it contains. Seen in its historic context, *Outrage* was the heir to two traditions, both of which are essential components in understanding the imagery that resulted. First, this monograph manifested a broad-based prejudice against suburbia that was a point of convergence for various different schools of planning and architectural thought. Second, even though it was not in itself a modernist tract in any strict sense, *Outrage* was party to a style of representation, a visual language, pioneered in Britain by the *Architectural Review*. That visual language had long since become part of the ideological framework within which the architects of the Modern Movement and their associates cast their ideas.

With regard to that imagery, we have focused on the specific medium of architectural photography to see something of how that imagery was created. In response to our original question of why a magazine that contained marvellous photographs of modern architecture also contained such indifferent pictures of suburbia, the answer would now seem relatively simple. It was that both types of photograph were brilliantly successful for the ends that the journal had in mind. The photography of the modern was heroic and exciting, the photography of suburbia effectively sapped any life from the depiction of the suburban environment. Seen in an historical context, the imagery of suburbia cannot be accurately interpreted without understanding its original juxtaposition with the imagery of heroic modernism. However, by the time that *Outrage* was produced, the illustrations of suburbia had become divorced from this other style of photography that helped to supply its original meaning. The visual language had been built up by an earlier generation; for much of the new post-war generation of readers, this was simply the way that suburbia was depicted. That style of depiction was perfectly commonplace and its deep-rooted place in the ideology of the Modern Movement ensured that it would be broadly accepted by those who subsequently espoused modernism.

Conclusion

Naturally, one would counsel caution before drawing too many broad conclusions from the particular case-study reviewed here. Certainly when dealing with suburbia it is possible to find various examples of sources that clearly incorporate the same style of lexical and visual imagery as *Outrage* and the *Architectural Review*, but, notwithstanding these, great care is needed before making inferences about how lasting an influence these publications have been on perceptions of suburbia. Indeed, such questions would inevitably founder in the conceptual mire that surrounds current debate about the role of imagery in the development of planning and architectural thought.[40] Nevertheless, there are reasons to suggest that the case-study in itself does raise issues that have more than just particularist, historical interest.

The first concerns the relationship between imagery and ideology. From the case-study considered here, we have seen the impress of ideology on the development of a view of suburbia, with an end-product that was as heavily filtered as some of the associated photography. Suburbia emerged not just as a housing type and a form of land use, but also as the antithesis of and, to some extent, obstacle to the Modern Movement's professional hopes and aspirations. With the contemporary retreat from modernism in the wake of the events of the last two decades, it is interesting to speculate how much the revival of interest in the characteristic visual imagery of suburbia – in the form of pseudo-vernacular design and traditional constructional materials – reflects the vicissitudes experienced by that ideology.[41]

Second, the imagery to be found in *Outrage* and similar publications was genuinely symptomatic of an approach to the depiction of suburbia that has existed for the last half-century or more. It remains the case when reviewing the broad ambit of urban studies that suburbia is either depicted in the sterile, unflattering, even deprecating manner that we have identified here or, quite simply, is not shown at all. It is a strategy that has rendered a sizeable chunk of the contemporary city as formless, ill-differentiated space rather than as recognizable, distinctive, and valued places. This is a point that carries considerable implications for the continuing business of interpreting the contemporary city.

The third, and last, point follows from this. At the start of this chapter, it was noted that there was a perceptible, if muted, movement towards reappraising the view of suburbia as being intrinsically deficient. In the hands of authors associated with this movement, one can find another, quite different reality ascribed to the suburban experience: of suburbs emerging as places of intrinsic interest, with infinitely varied building styles, considerable scope for expression of individuality, and possessing recognizable, and often warm, community relations.

It is in no way part of the remit of this chapter to argue that such viewpoints are automatically more palatable than the perspective of suburbs as being inherently deficient; merely that such views are equally plausible on *a priori* grounds and deserve to be afforded no more or less credence, in the absence of evidence, than views to the contrary. In arguing this, there is an interesting parallel that can be drawn with the reappraisal of the dominant attitudes towards the *inner* city of 25–30 years ago. During the 1950s and early 1960s, a period marked by an emphasis on urban clearance and comprehensive redevelopment, a series of studies[42] was undertaken by social scientists in run-down areas of working-class housing in the inner city. The wealth of data that these studies produced challenged popular and professional views of the inner city as being invariably socially as well as physically blighted and, in the process, threw a sustained and critical focus upon the behavioural environment of decision-makers. Gradually, it was realized that the view of the 'insider' (the resident) might well be very different from that of the 'outsider' (the urbanist, the expert, the 'planner') and that there might be factors, not readily apparent to the casual observer, that could offset the disadvantages of the blighted physical environment. Such perspectives, in turn, contributed in no small measure to the important reappraisal of the policy objectives that took place in the late 1960s and early 1970s.

The physical quality of the suburban environment is normally quite different, of course, from that of the inner city. Nevertheless, one wonders whether it is really too far fetched to suggest that, in the 1980s, we might absorb something of the same message and make a serious start on reappraising the view that the *outer* city is visually, socially, and morally impoverished.

Acknowledgements

John Gold would like to express his gratitude to the Department of Geography of the London School of Economics for generously affording him facilities and technical support as an Academic Visitor in 1986–7. This paper was written during this period. We would also like to thank Ms Shirley Hind of the Architectural Press for her assistance in obtaining photographs that accompany this article and to the Architectural Press for their kind permission to allow them to be republished here.

Notes

1 M. Robertson, *Laymen and the New Architecture* (Murray, London, 1925), p. 68.
2 P. Abercrombie, 'Introduction', in P. Abercrombie (ed.), *The Book of the Modern House: A Panoramic Survey of Contemporary Domestic Design* (Waverley, London, 1939), p. xix.
3 I. Nairn, 'Gentle, not genteel', *Architectural Review*, vol. 130 (1961), p. 387.
4 I. Nairn, 'Spec.-built', *Architectural Review*, vol. 129 (1961), p. 164.
5 Abercrombie, 'Introduction', in Abercrombie (ed.), *The Book of the Modern House*, p. xix.
6 R. Sinclair, *Metropolitan Man* (Allen & Unwin, London, 1937), p. 106.
7 T. Sharp, 'The English tradition in the town. III: Universal suburbia', *Architectural Review*, vol. 79 (1936), pp. 115–20.
8 D. Burtenshaw, M. Bateman, and G. J. Ashworth, *The City in Western Europe* (John Wiley, Chichester, 1981), p. 304.

9 Nairn, 'Spec.-built', p. 164.
10 L. Mumford, *The City in History* (Penguin, Harmondsworth, 1963), p. 584.
11 R. Abler, D. Janelle, A. Philbrick, and J. Sommer (eds), *Human Geography in a Shrinking World* (Duxbury Press, North Scituate, Mass., 1975), p. 71.
12 P. Oliver, 'Round the houses', in A. C. Papadakis (ed.), *British Architecture* (Academy Editions, London, 1982), p. 17.
13 For example, C. Cooper Marcus and L. Hogue, 'Design guidelines for high-rise housing', *Journal of Architectural Research*, vol. 5 (1976), pp. 34–49; W. Michelson, *Environmental Choice, Human Behavior and Residential Satisfaction* (Oxford University Press, New York, 1977).
14 J. M. Richards, *The Castles on the Ground* (Murray, London, 1946); H. Gans, *The Levittowners* (Pantheon, New York, 1967); A. A. Jackson, *Semi-detached London* (Allen & Unwin, London, 1973); A. M. Edwards, *The Design of Suburbia* (Pembridge Press, London, 1981); P. Oliver, I. Davis, and I. Bentley, *Dunroamin* (Barrie & Jenkins, London, 1981); E. Relph, *Rational Landscapes and Humanistic Geography* (Croom Helm, London, 1981), pp. 84–105; D. N. Rothblatt and D. J. Carr, *Suburbia: An International Assessment* (Croom Helm, London, 1986).
15 W. Kirk, 'Problems of geography', *Geography*, vol. 48 (1963), p. 366.
16 I. Nairn, 'Outrage', *Architectural Review* (special issue), vol. 117 (1955), pp. 361–460; republished as *Outrage* (Architectural Press, London, 1956).
17 I. Nairn, *Counter-attack against Subtopia* (Architectural Press, London, 1956). For insight into Nairn's thought and writings, see I. Nairn, *Your England Revisited* (Hutchinson, London, 1964), *The American Landscape* (Random House, New York, 1965), and 'Through a glass darkly: Outrage 20 years after', *Architectural Review*, vol. 158 (1975), pp. 328–37.
18 G. Stapledon, *The Land Now and Tomorrow* (Faber & Faber, London, 1944).
19 Nairn, 'Outrage', p. 363.
20 Ibid., p. 365.
21 Ibid., p. 372.
22 Ibid., p. 367.
23 Ibid., p. 368.
24 For a general overview of the uses and abuses of architectural photography, see T. Picton, 'The craven image, or the apotheosis of the architectural photograph', parts 1 and 2, *Architects' Journal*, vol. 170, no. 30 (25 July 1979), pp. 175–90, and vol. 170, no. 31 (1 August 1979), pp. 225–42. It should also be stressed that certain of the features of this style of photography had already been pioneered in Germany during the 1920s, a point made by S. Bayley, 'The roaring tendencies', broadcast script, BBC Radio 3, 27 October 1977.
25 J. M. Richards, *Memoirs of an Unjust Fella* (Weidenfeld & Nicolson, London, 1980), p. 137.

26 Le Corbusier, *The Athens Charter* (Lund Humphries, London, 1973),
   p. 61 (originally published 1943 as *Le Charte d'Athènes*).
27 R. Barthes, *Mythologies* (Jonathan Cape, London, 1972).
28 For fuller discussion of the notions of image ambiguity and 'anchorage'
   which have been touched on here, see F. Webster, *The New
   Photography* (Calder, London, 1980), ch. 6.
29 For example, see A. W. Gouldner, *The Dialectic of Ideology and
   Technology* (Macmillan, London, 1976); D. Bell, 'Ideology', in A.
   Bullock and O. Stallybrass (eds), *The Fontana Dictionary of Modern
   Thought* (Fontana, London, 1977), pp. 298–9; J. Larrain, *Marxism
   and Ideology* (Macmillan, London, 1983).
30 D. J. Manning and T. J. Robinson, *The Place of Ideology in Political
   Life* (Croom Helm, London, 1985), p. 1.
31 On the Modern Movement generally, see R. Banham, *Theory and
   Design in the First Machine Age* (Architectural Press, London, 1960);
   L. Benevolo, *History of Modern Architecture*, 2 vols (Routledge &
   Kegan Paul, London, 1971); K. Frampton, *Modern Architecture: A
   Critical History* (Thames & Hudson, London, 1980); W. R. Curtis,
   Modern Architecture Since 1900 (Phaidon Press, Oxford, 1982); V. M.
   Lampugnani, *Architecture and City Planning in the Twentieth Century*
   (Van Nostrand Reinhold, New York, 1985).
32 A. Jackson, *The Politics of Architecture* (Architectural Press, London,
   1970), p. 28.
33 Material relating to modernism and to interviews with leading figures
   in the English Modern Movement is derived from work carried out by
   J. R. Gold in connection with a forthcoming monograph entitled
   'Modernism and the urban imagination'.
34 M. M. Gold, '"A place of delightful prospects": promotional
   literature and the imagery of suburbia', paper presented to 25th
   Congress, International Geographical Union, Paris, 1984.
35 Jackson, *Semi-detached London*, p. 205.
36 Interview by J. R. Gold with J. M. Richards, 3 December 1986. See
   Note 33.
37 For more details on this topic, see: J. R. Gold, 'The death of the urban
   vision?', *Futures*. vol. 16 (1984), pp. 372–81; and J. R. Gold, 'A world
   of organised ease: the role of leisure in Le Corbusier's *La Ville
   radieuse*', *Leisure Studies*, vol. 4 (1985), pp. 101–10.
38 See Note 36.
39 Interview by J. R. Gold with P. E. A. Johnson-Marshall, 18 December
   1986. See Note 33.
40 Gold, 'The death of the urban vision?'; J. R. Gold, 'The city of the
   future and the future of the city', in R. King (ed.), *Geographical
   Futures* (Geographical Association, Sheffield, 1985), pp. 92–101.
41 An interesting discussion of this point can be found in S. Lyall, *The
   State of British Architecture* (Architectural Press, London, 1980),
   pp. 70–93.
42 See discussion in J. R. Gold, *An Introduction to Behavioural
   Geography* (Oxford University Press, Oxford 1980), pp. 164–74.

Chapter ten

Divided perception in a united city: the case of Jerusalem

Michael Romann

Introduction

Cognitive maps of urban space

The problem of understanding how individuals and groups make sense of their environments has been a theme that has surfaced in a range of social and urban geographical contexts. Studies of the urban mental maps of city residents reveal how limited and selective such cognitive images are, and how they diverge from the city's objective geographical reality. This disjunction between environments as they are, and as they are taken to be, was of course at the heart of William Kirk's behavioural environment model. Here, he isolated the perceived world as a distinct entity crucial for understanding both decision-making processes and the spatial behaviour of individuals and groups alike. Kirk's schema proved to be widely applicable not only to historical geography, for which it was originally conceived, but also to the elucidation of the role of perception as a factor in contemporary social-geographic behaviour.

In contemporary urban society the behavioural environments of residents are constituted not only by the perceptual filters through which individuals and communities structure the world, but also by the simple fact that those individuals and communities possess only selective familiarity with particular areas and specific locations. Empirical research emphasizes two additional points of interest. First, the extent of residents' familiarity with the urban environment principally reflects daily activity patterns. Second, differences in perception among groups within a city reflect differential socio-economic status. Groups with higher income, education, and mobility enjoy greater familiarity with the urban environment than those of the lower rungs of the socio-economic ladder, with minority ethnic groups being particularly limited in their knowledge. By the same token large host communities have little experience of minority ghetto areas and thus both groups live, as it were, in more or less hermetically sealed separate worlds. Urban familiarity, it seems, reflects patterns of everyday experience, socio-economic status, ethnic affiliation, and majority–minority group interaction.[1]

The particular case of Jerusalem

Jewish and Arab familiarity with Jerusalem as a whole, and with each of its two sectors in particular, may serve as an important indication of the pattern of mutual relations – both functional and spatial – which has developed between the two population groups. In Jerusalem's case, spatial familiarity is of particular significance, given the almost total residential, commercial, and public-service segregation that has persisted between Jews and Arabs even in the administratively united city. Consequently, intersectoral relationships necessarily involve crossing over to the 'other side': accordingly Jews and Arabs have acquired only selective information of each other's sector. With Jerusalem, urban cognitive differences between Jews and Arabs arise not only from differential socio-economic status but also – and particularly – from national and ethnic identities. Cultural differentiation and political conflict are factors which influence – both directly and indirectly – people's daily behaviour as well as the nature of the structural relationships between the Jewish majority and the Arab minority. In addition we expect that a parallel examination of Jewish and Arab spatial awareness will provide insights on other issues too, particularly the extent to which Jerusalem is actually perceived as a united city by the two groups who live within its confines.

A number of specific questions arise out of such a study. First, to what extent does an 'ethnic boundary' continue to exist in Jerusalem? Second, is either of the two groups the more knowledgeable of the other's space and if so how and why is this the case? A third question concerns whether the Jewish and Arab selective awareness of Jerusalem that may exist reflects, at one level, patterns of daily behaviour, and at a deeper level structural processes of interethnic reliance.

Before attempting to answer these questions it is necessary to set the scene. Even before 1948, religious, cultural, and national differences between Jews and Arabs had already occasioned widespread segregation in daily life. Members of both communities tended to live in well defined residential quarters and developed a whole range of separate community services. Daily interactions in the labour and commercial markets also gradually decreased over time following periodic eruptions of intercommunal violence. The physical division of Jerusalem between Israeli West Jerusalem and Jordanian East Jerusalem as a consequence of the 1948 War brought about not only a complete separation between Jews and Arabs but the creation of totally unconnected social and political urban units. Since the armistice line ran right through the middle of the city, two parallel commercial centres along with various other infrastructural systems had to be developed away from the hostile border. Population growth and economic development also took place in an unrelated fashion, reflecting the different conditions on each side of the political boundary.

Following the Six Day War in 1967, Jerusalem was physically and administratively reunited under direct and exclusive Israeli rule. In order to enforce its political control and prevent the future repartition of the city, Israel took two major steps. First, it extended its legal jurisdiction over East Jerusalem and the surrounding suburbs with the result that within the newly extended municipal boundaries the Jewish population maintained a substantial majority. Second, a massive programme of Jewish settlement beyond the former partition line was undertaken: the former Jewish Quarter in the Old City was re-occupied and major new Jewish residential neighbourhoods and activity centres were established all over East Jerusalem (Figure 10.1). By the mid-1980s close to one-third of the entire Jewish population of the city now lived in East Jerusalem, a figure approaching that of the total local Arab inhabitants. The removal of the boundary between the two city parts was further expressed in the emergence of extensive intersectoral relationships, particularly in the economic sphere. Thus about 40 per cent of the entire Arab labour force became regularly employed in the Jewish sector, primarily in low-status manual occupations. Jews and Arabs re-established a wide network of exchanges between the two business communities and their respective customers.

However, significant ethnic divisions persisted, mirroring pre-unification realities as well as ongoing social distinctions and political antagonisms. West Jerusalem remained entirely Jewish, while in East Jerusalem, although Arab and Jewish residential areas developed close to each other in a kind of mosaic pattern, residential segregation on a neighbourhood level remained almost total.[2] The Jewish and Arab sectors maintained their separate, distinct business districts, public transportation systems, and a whole range of other parallel public services from educational and medical facilities to professional organizations. Fundamentally, all

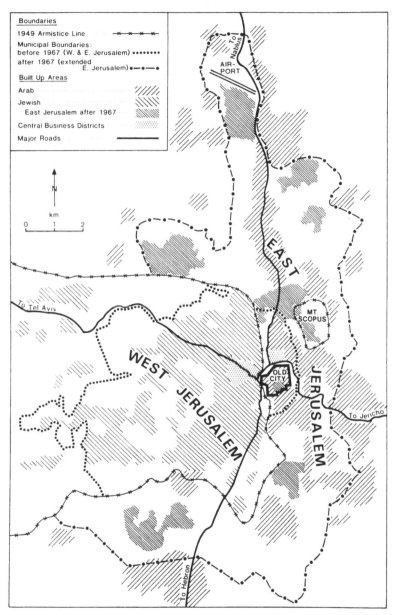

Figure 10.1 Jerusalem: important boundaries and the pattern of the
built-up area

of the urban functions and urban space could be classified as either 'Jewish' or 'Arab'. Indeed, in post-1967 Jerusalem, ethnic boundaries – 'divisions in the mind' – remain highly significant for everyday conduct.[3]

Methodology

In an attempt to clarify some of these issues, a survey was carried out in both sectors of the city in 1982 amongst Jewish and Arab high-school students,[4] who were requested to fill out a questionnaire that was identical save for the fact that one was in Hebrew and the other in Arabic. The questionnaire consisted of two parts. In the first section students were requested to locate listed sites selected from throughout Jerusalem, including neighbourhoods and streets, public institutions, and principal commercial establishments (see Figure 10.2). Certain key sites with parallel functional significance in Jewish West Jerusalem and the Arab section of East Jerusalem were chosen. Students were asked if they knew where the site was and whether they had ever visited it, and then to describe its location briefly in writing. This procedure enabled respondents to be classified into three groups: those who knew where the site was and who could provide an accurate description; those who did not know where it was; and those who thought they knew where it was but whose descriptions revealed that they evidently did not. Visits to the 'other side' of Jerusalem were the subject of the second part of the student questionnaire. Here, questions were designed both to gain some basic information about the patterns of visiting and also to begin to understand aspects of the psychology of the 'crossing-over' experience by examining more subjective states of consciousness. Accordingly, as well as asking about the purpose and frequency of visits, questions were also included about feelings of comfort or discomfort, whether crossing the ethnic boundary occurred alone or in company, and about 'interest' in the opposite sector.

To be sure, the survey population is fully representative neither of the school nor the total population of the city. The sample was limited principally to students within the general municipal stream because it was easier to gain access to these than to Jewish religious schools and Arab private ones. It must also be remembered that these students are drawn from a demographic group which has spent nearly its entire life within a physically united Jerusalem. Furthermore, school children are not yet as economically active as adults and are therefore less likely to make use of the other sector.

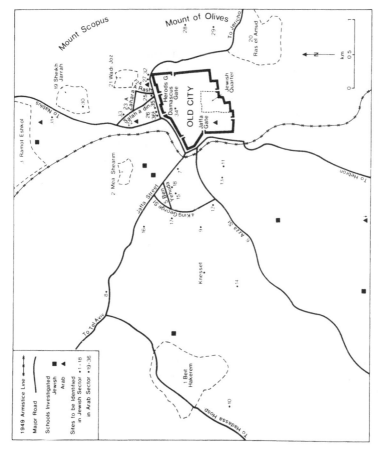

Figure 10.2 Jerusalem: schools surveyed, sites to be identified in the survey, and important locations referred to in the text (see Table 10.1 for key to site numbers)

Patterns of urban familiarity

It is, of course, not at all surprising that Jewish and Arab students are more able – in fact over three times more able – to identify sites in their own sector of the city than can members of the other group (see Table 10.1). To be sure, this does not mean that Jews are entirely ignorant of Arab East Jerusalem or Arabs of West Jerusalem. What is particularly instructive, however, is the range of sites in the opposite sector with which each group is familiar – a number of nearby and well-known residential areas, the main commercial arteries in both sides of the city, the main bus terminals, central post offices, and a principal hotel in each sector. Certainly, some Jewish and Arab students were also able to identify a range of additional sites on the 'other side', such as distant residential areas, secondary streets, hospitals, and cinemas, but this ability was far less common amongst members of both groups. This is illustrated by the fact that while 80 per cent of both Jews and Arabs were able to locate one of the well-known high schools in their own sector, only 4 per cent of them were able to do the same for a similar school in the other.

Right from the outset, then, it is plain that an ethnic boundary runs through the heart of Jerusalem; and this is further attested by the simple fact that geographical proximity of respondents' schools or places of residence to the 'other side' brings with it no greater knowledge. In Jerusalem proximity does not mean familiarity. Take, for example, Jewish and Arab students in two nearby schools within eyesight of each other; their knowledge of the 'other side' is scarcely better than that of students from schools far further away. Here, geographical distance has little bearing on cognitive distance. The same is also true of residential areas: Jews residing close to the Arab area of the city (either in nearby West Jerusalem or in the new Jewish neighbourhoods established after 1967 beyond the old partition line) were hardly better able to identify Arab locations than those living miles away in neighbourhoods well within West Jerusalem. As for the Arab students, geographical propinquity is, if anything, even less important: those living relatively close to the western sector had as little knowledge of the Jewish side as those in villages at the city's periphery or even beyond the municipal boundary.

For both Jews and Arabs, sites in their own sector are always perceptually 'closer' than those in the other. As Kirk's behavioural model implies, there is an important distinction to be drawn between objective geographical distance and subjective perceived distance. Nowhere is this more clearly exemplified than in the fact

Table 10.1 Site familiarity (Jewish and Arab students in Jerusalem)[a]

Sites in the Jewish area	Jews	Arabs	Sites in the Arab area	Arabs	Jews
	%	%		%	%
Neighbourhoods and streets					
1. Beit Hakerem	89.7	14.4	19. Sheikh Jarrah	82.4	21.4
2. Me'a She'arim	83.5	10.5	20. Ras el Amud	82.3	7.9
3. Ramot Eshkol	86.6	32.0	21. Wadi Joz	73.2	50.0
4. King George Street	96.9	41.8	22. Salah a Din Street	83.1	51.9
5. Ben Yehuda Street	93.3	34.6	23. A Zahara Street	64.7	3.0
6. Azza Street	67.1	5.2	24. A Rashid Street	32.0	2.4
Institutions and public buildings					
7. Central Post Office (West Jerusalem)	63.4	43.8	25. Central Post Office (East Jerusalem)	78.5	41.4
8. Central Bus Terminal (West Jerusalem)	79.3	39.9	26. Central Bus Terminal (East Jerusalem)	60.8	39.6
9. Rehavya Gymnasium	77.4	4.0	27. A Rashidiya Gymnasium	81.7	4.8
10. Sha'arei Zedek Hospital	92.7	19.0	28. El Mukased Hospital	83.7	2.4
11. King David Hotel	90.0	36.6	29. Intercontinental Hotel	64.7	39.6
12. Kings Hotel	81.7	19.6	30. Ambassador Hotel	41.2	12.8
Places of entertainment and commercial sites					
13. Main Football Field (West Jerusalem)	60.3	20.3	31. Main Football Field (East Jerusalem)	53.6	9.1
14. Israel Museum	95.7	32.1	32. Rockefeller/Palestine Museum	38.0	54.8
15. Orion Cinema	92.7	19.7	33. Al Hamra Cinema	80.4	15.9
16. Mahane Yehuda Market	86.5	38.6	34. Khan a Zeit Market	72.6	3.7
17. Hamashbir Department Store	97.0	48.4	35. Nasser a Din Supermarket	53.0	0.6
18. Atara Café	76.2	7.9	36. A Sha'ab Café	43.1	0.0
All sites (average)	83.9	26.0		64.9	20.1

Note: a. The students' 'familiarity' with sites is based on their responses to the question: 'Do you know where this place is?', and their written responses concerning the location of the various sites. A 'familiarity' score was given to those students who described where the place was, even if their description was approximate. This excluded students who noted categorically that they were not familiar with the site, who did not respond to the question, and those whose identification of the site's location was clearly incorrect.

that the Rockefeller Museum was the site in Arab Jerusalem most commonly identified by Jewish students. Since this building in fact became a 'Jewish institution' following 1967 it is not surprising that Jewish students were more familiar with its location than the Arabs were.[5] As far as mental images are concerned, the Rockefeller Museum 'belongs' more to Jewish West than to Arab East Jerusalem irrespective of its physical location.

That there are ethnic factors at work in the construction of Jewish and Arab mental geographies is further revealed by the fact that Arabs seem to have more familiarity with the Jewish sections of the city than the converse. Certainly the difference is small – of the order of 6 per cent. Still, even this is highly significant when we take into account the fact that Arab students were less able correctly to locate sites in their *own* area of the city. By the use of a weighting procedure designed to incorporate this finding,[6] the gap dramatically widens – from 6 to 16 per cent – Arab abilities to locate sites in the Jewish sector being nearly 70 per cent higher than Jewish abilities in the Arab. This manifests itself in a number of ways. A significant number of Arab students, for example, are familiar with all three main streets in the West Jerusalem commercial centre, being able to locate Ben Yehuda and King George Streets in relation to Jaffa Street, whereas most Jewish students were only able to identify one main street in the commercial centre of East Jerusalem – Salah a Din – while largely ignoring the neighbouring two main commercial arteries, A Zahara and A Rashid Streets. Besides this a number of Arabs could also pinpoint in the west of the city the central vegetable market, one of the principal department stores, a well-known coffee-house, and a large hospital; by contrast hardly any Jewish students could identify parallel sites in Arab East Jerusalem. Thus, whereas the Jewish mental geography of the Arab side has only a few landmarks – tourist sites and the major commercial artery – the Arabs have a greater awareness of the Jewish urban terrain.

The meaning of urban familiarity

Having outlined the patterns of urban cognition in the Jerusalem context, it is important now to pause and reflect on what people mean by 'familiarity' and what the relationship is between the *subjective feeling* of being familiar with a site and *objective knowledge* of its location. The concept of 'familiarity' is understood in different ways. For one thing, it is very common for people to *say* they are familiar with some site or other; yet in our particular survey many among these self-same respondents explicitly stated that they did not know where the site on the 'other side' was located and sometimes added that they had never even visited the opposite sector! No doubt their claim to be familiar with certain parts of the city has often something to do with self-perception or with the expectations of others. Besides, many who claimed to be familiar with a particular location, even in their own sector, were just simply unable to locate it correctly.

There are then difficulties in understanding what respondents themselves mean by 'familiarity'. Beyond that, however, there are also problems of determining from questionnaire answers whether respondents actually *are* familiar with a site or not. Take the disarmingly simple question: 'Where is this site?' Interpreting responses turns out to be notoriously difficult. For a start many of the students did not answer this question at all; and then those who did make a stab at an answer often provided brief descriptions which were, to say the least, vague.[7] The question 'Where is the central post office in West Jerusalem?', for example, produced such answers as 'at the beginning of Jaffa Street', 'at the end of Jaffa Street', 'by Jaffa Street', 'opposite Jaffa Gate', 'in the centre of the city'. From these it is rather difficult to deduce which students really do know where the central post office is. Nevertheless, taken overall, a rudimentary classification of answers into three broad categories – 'correct', 'approximate', or 'incorrect' – was sufficient to make sense of the data.

Table 10.2 Urban familiarity: perceived and objective (Jewish and Arab students in Jerusalem)[a]

	Sites in the Jewish area		Sites in the Arab area	
	Jews	*Arabs*	*Arabs*	*Jews*
	%	%	%	%
Responses[b]				
Familiar with				
(from those who responded)	93.4	49.8	85.1	35.7
(from total survey population)	90.5	42.2	80.7	32.1
Visited the site				
(from those who responded)	89.6	40.2	77.4	27.6
(from total survey population)	86.5	32.9	72.6	24.1
Identification of site[c]				
Did not specify	13.8	71.1	32.4	73.9
Incorrect specification	2.3	2.9	2.7	6.0
Approximate specification	10.3	7.4	14.5	7.5
Correct specification	73.6	18.6	50.4	12.6
(Approximate & correct				
specification)	(83.9)	(26.0)	(64.9)	(20.1)

Notes: a. Average scores for a range of eighteen sites in each area (see Table 10.1).
     b. *Relates to student responses (yes or no) to the questions: 'Do you know where the place is?' and 'Have you ever visited the site?'*
     c. Relates to the written responses concerning a description of the site's correct location.

The differences between claims to familiarity, actual visits, and ability to locate sites are statistically displayed in Table 10.2. Particularly arresting are the ethnically related familiarity characteristics revealed in the selective knowledge which both Jewish and

Arab students possess of the two sides of the city. Understandably, both groups have higher familiarity, visit, and identification ratings for their own 'sectoral side'; and yet, compared with Jews, Arab familiarity with their own area and ability correctly to locate sites therein were systematically lower. Certainly there was little difference between the groups in terms of *claims* to familiarity, but when actual abilities at locational specification were compared, significant differences surfaced. For some reason, Jewish students have a better handle on their own urban geography than do Arab students on theirs. Still, for all the apparently greater urban awareness of the Jews, it is clear that Arab students have greater knowledge of the Jewish area than do Jews of the Arab.

The relative ignorance of the geography of the 'other side' can be attributed to the dynamics of a culturally and politically 'divided space'. Site identification is, of course, closely bound up with naming, and the different names that are commonly used by each population group when referring to the same site produces its own problems. A couple of examples are illustrative. The Rockefeller Museum is known as the Palestine Museum by the Arab population (and was accordingly translated as such in the Arab questionnaire) and this fact doubtless aided some Arabs in their identification of it. By the same token, the absence of a Hebrew name for the Khan a Zeit market (the main market in the Old City, beginning at the Damascus Gate) meant that hardly any Jewish students were able to locate it, despite the fact that most, if not all, of them had often visited it. And the problem is more widespread than just this: the different names of the gates of the Old City (for example, what is the Jaffa Gate for Jews is the Bab el-Khalil or Hebron Gate for Arabs) is only symptomatic of a whole range of different linguistic labels employed for identifying key sites. What is important in this context is that the continued existence of an ethnic perceptual boundary between Jews and Arabs in Jerusalem, reflected in the choice of naming practices, results in a geography of limited mutual awareness.

The existence of a blurred perceptual map of the 'other side' of the city is also evident from some written responses by the students. The absence of sufficient points of reference, and the tendency to 'reduce the distances' between them illustrates this. Thus, for example, many of the Jewish students noted that Salah a Din Street is located opposite the Damascus Gate, when in fact it begins at Herod's Gate. Similarly, Arab students observed that the Beit Hakerem neighbourhood was to be found 'next to the Hadassa hospital' or that the King David Hotel was located 'opposite the Jaffa Gate', descriptions which would amaze Jewish observers due to the great distances involved. Indeed, the way in which places were related to each other and the different descriptive terms used by Arabs and Jews basically reflected the different observational stances of each group. The 'centre of the city' in Jewish terminology is 'Jaffa Street' for the Arabs, what is 'before' for Arabs is generally 'beyond' for Jews. Accordingly, accurate identification was obviously easier for both lroups when sites were near well-known and prominent co-ordinates such as the Mount of Olives or the Intercontinental Hotel, the Knesset, or the Israel Museum.

In understanding perceptual geographies, error is often as enlightening as accuracy. Many of the Jewish students located the main Arab bus terminal next to Jaffa Gate – precisely the same spot that some Arabs specified as the site of the main Jewish bus terminal! In part this erroneous identification may be attributed to the fact that there is a concentration of both Arab and Jewish buses and associated parking spaces at this site. Doubtless the fact that this area is one of the principal meeting places for the two populations is also an important factor here. Also instructive is the fact that many of the Jewish respondents actually located the East Jerusalem main bus terminal, post office, and football field within West Jerusalem. Evidently, many Jewish students practically ignore the existence of the Arab section of the city altogether.

Finally, Arabs and Jews have rather different ideas about what constitutes the general urban area – namely, what is included within Jerusalem and what is not. This was especially apparent in the responses of Arab students who live in peripherally located suburbs and villages in the east of the city which were annexed by Israel and incorporated within the extended municipal boundaries after 1967. These students often defined sites in the Arab sector of the city as being located 'in Jerusalem', as though they were referring to a different area from that in which they themselves currently lived. In other words, while in Jewish perceptions the Jerusalem municipal area includes the whole of the eastern section of the city annexed by Israel, this is rarely the case for the Arab population. The Arab population, it seems, regard 'their' Jerusalem as defined by the restricted jurisdictional limits that had existed before 1967. Besides, the Arabs – unlike the Jewish residents of the city – do not differentiate between the new Jewish neighbourhoods that have come into being since 1967 within the east of the city and those which have been established beyond the city limits in the West Bank. In Arab eyes, both types are viewed as '*hitnachalut*', that is, illegal settlement of Arab territory. The distinction made between the annexed area of East Jerusalem on the one hand and the general West Bank region on the other, so it appears, is one that is only made by the Jewish population.

Moving across the ethnic boundary

One means of getting to grips with the significance of the ethnic boundary that has been shown to run through Jerusalem is to focus on the movement of people between the two sides of the divide. This certainly involves, as has been attempted above, establishing the patterns of intersectoral visitation; but, more important, it requires investigating the underlying reasons for making these trips. Indeed, what may be even more significant for understanding just how deep ethnic divisions run are the feelings that people have and the circumstances under which they make the transition between East and West Jerusalem. Accordingly some attempt will be made to get into the minds of members of each group in order to begin to construct what might be called the geography of mutual uncertainty.

The purpose of visits

From the Jewish perspective, visits to East Jerusalem were primarily either for general outings or to visit historical and religious sites (Table 10.3). However, there were still a substantial number whose purpose in crossing over was for shopping, often associated with visits to the Old City and its markets. Besides these, some Jewish students explicitly mentioned more specific reasons such as visiting an Arab cinema and this was verified by their ability correctly to locate a major East Jerusalem cinema. On the other hand relatively fewer Arab students penetrated into Jewish territory for general outings or shopping trips, but relatively more did so in order to go to the cinemas, for employment, or to make use of the swimming pools, the municipal zoo, and city parks. Compared to the Jews, then, Arab excursions into the 'other side' are more varied and more specifically targeted. Thus, when Jews were asked about their most recent visit to the 'other side', general outings, visiting historical sites, or shopping were by far the most frequently reported purposes; by contrast, the Arab figures are more evenly distributed among 'shopping', 'general outing', 'work', 'administrative arrangements', or 'other purposes'.

All this goes to show a much greater and more diverse Arab reliance on Jewish West Jerusalem than the reverse, and clearly reflects the patterns of mutual need and exchange between the Jewish and Arab populations in general. This relationship is

Table 10.3 Purpose of visits (Jewish and Arab students in Jerusalem)[a]

Purpose of visit	Jewish visit to Arab area	Arab visit to Jewish area
	%	%
General outing	88.4	38.9
Historical or religious site	82.9	—[b]
Museum	43.9	34.6
Restaurant	37.2	26.1
Café, club	21.3	
Cinema	12.2	28.8
Public park, zoo	—[b]	52.9
Swimming pool	—[b]	24.2
Shopping	62.6	34.6
Employment	5.5	23.5
Studies	32.9	11.8
Visiting friends	12.2	17.6
Administrative arrangements	3.7	11.1
Other purposes	6.7	19.6

Notes: a. Based on positive responses to the question: 'With respect to all your visits to the Arab/Jewish [as appropriate] section of the city, have you ever made the visit for one of the following purposes?'
b. In these cases, the question was not put because of the absence of such options.

perhaps best described in terms of complementarity: each sector is able to offer the other certain goods or services not readily available on their own side. Take, for example, the fact that on the Arab side the tourist sites and the markets of the Old City are open during the Jewish Sabbath whereas everything is closed in the Jewish sector; conversely, the large retail outlets, and a whole range of public-sector services – from medical clinics to swimming pools – constitute major magnets in the Jewish sector. Indeed, the social characteristics and behavioural norms of these two highly different ethnic groups are themselves complementarity factors. Thus, many Arab young people visit West Jerusalem cafés and cinemas simply because there they can meet Arabs of the opposite sex – something for which there is limited opportunity in their own part of the city.

Having established that each of the two sectors of the city supply commodities to the other side, it must not be assumed that the *balance* of mutual transaction is symmetrical. On the contrary, because the Jewish sector is both larger and more commercially developed, the Arab population is far more dependent on economic opportunities offered there than the reverse. Employment opportunity is, of course, a particularly marked expression of this dependency relationship, as are the range of specialized services requiring a high threshold population and the various state and municipal functions, all of which are located in the west of the city. Indeed, it would be true to say that, while the Arab population *has to* use the Jewish sector because it has no other choice, the reverse is rarely true: Jews do not have to use the Arab sector even if at times it is more convenient to do so. These economic relationships, moreover, invariably have spatial expression. Consider the organization of public transport in Jerusalem. The Jewish bus network serves virtually every Jewish destination throughout the city, including those in distinctly Arab areas; the Arabs on the other hand have to resort to the Jewish bus service the moment they cross to the west of the city, for want of any alternative.

The different mental maps that Jews and Arabs have of Jerusalem are largely reflective of the functional relationships between the two ethnic sectors of the city. Selective mutual dependency is mirrored in differential urban familiarity: Arab students have greater knowledge of the Jewish sector simply because they are often *obliged* to visit the west of the city. In addition, the overall patterns of interaction are also reflected in the selective familiarity with specific sites: both Jewish and Arab students were more familiar with the retail centres in both sectors; knowledge of hospital location in the opposite sector is confined to Arabs; while neither of the populations are familiar with the parallel segregated schools.

What further helps to explain the overall limited familiarity with the 'other side' are the persisting, widespread patterns of institutional, functional, and spatial segregation. There is, for example, a signal lack of indirect information about what takes place beyond each group's own sector. Arab and Jewish local newspapers give only very limited coverage to daily affairs on the 'other side', and the same is true of commercial advertising, cinema publicity, pharmacy rotas, and individual sales and purchase adverts. The Israeli authorities have published neither a telephone directory nor city map in Arabic, while Arab entrepreneurs have published their own independent telephone directory which virtually ignores the Jewish subscribers.

The patterns of familiarity with the 'other side' are evidently both cause and consequence of the lack of social interaction. Not only do these urban mental maps limit the range of interethnic relationships, but they are themselves the embodiment of the social history of ethnic separation.

Behavioural characteristics

While all these findings clearly indicate the persistence of an ethnic boundary within Jerusalem, the real presence of this invisible 'barrier' is revealed in certain aspects of what could be called the psychology of 'crossing-over behaviour'. Initially we should note some basic facts about the extent of trans-Jerusalem visiting (Table 10.4). While two-thirds of the Jewish students stated that they had visited the 'other side' in the month preceding the survey, the parallel rate amongst Arab respondents was around 50 per cent. On the surface this degree of movement might seem to run counter to the idea of an ethnic boundary dividing the city; and yet even in this volume of visiting a hint of the persistence of the ethnic factor is clearly discernible, for Jews more frequently penetrate into the Arab zone than vice versa. Thus, while all of the Jewish respondents had visited the Arab sector on some occasion, 12 per cent of the Arab male students and 17 per cent of the Arab females had never crossed to the 'other side'.

A major psychological manifestation of the intrusion of an ethnic boundary is the feeling of discomfort that groups experience as they cross into alien territory. The passage into foreign territory, as in the case of Jerusalem, is typically signalled by contrasting architectural styles, different days of closing, and shop signs in different languages on either side of the line. It is therefore not surprising that in Jerusalem a substantial proportion of both Jews and Arabs spoke of feeling uncomfortable when visiting the other sector. What is particularly arresting in this context is the fact that the Arab males are the group which is most likely to feel uneasy. Doubtless this reflects the realities of the more general geography of internal movement within Jerusalem. Many Arab males have to travel alone into the Jewish sector because their work takes them across the boundary and thus they frequently find themselves as isolated passengers on a Jewish bus, as lone workers in a shop or factory, or as maintenance personnel in some remote residential neighbourhood in the west of the city. Jews, by contrast, are generally able to choose when and how to cross over and thus they are seldom exposed to comparable situations. If the solitary excursion into Jewish territory brings with it a sense of psychological insecurity for Arabs, the corollary is that those, whether Jews or Arabs, who feel the need for accompaniment as they cross to the other side often do so in order to relieve some of the anxiety and tension that they may feel. The fact is that the majority of the youth who visit the opposite sector of Jerusalem do so in groups. Here, internal psychological states find expression in observable forms of movement.

Table 10.4   Visiting behaviours (Jewish and Arab students in Jerusalem)

	Jews in Arab areas		Arabs in Jewish areas	
	Males %	Females %	Males %	Females %
Most recent visit[a]				
During the last month	67.1	66.5	56.6	49.1
Between a month and a year ago	28.6	30.4	13.1	15.1
Over a year ago	2.9	3.1	17.2	11.3
Have never visited	—	—	12.1	17.0
Did not answer	1.4	—	1.0	7.5
Feeling of ease[b]				
Feel totally at ease	21.4	12.8	20.2	26.4
Feel reasonably at ease	48.6	48.9	16.2	32.1
Feel slightly uneasy	28.6	37.2	25.3	13.2
Feel totally uneasy	1.4	1.1	35.3	17.0
Did not answer	—	—	3.0	11.3
Accompaniment on the visit[c]				
Always go alone	1.4	—	10.1	5.7
Usually go alone	11.4	2.1	24.2	9.4
Seldom go alone	44.3	23.4	36.4	26.4
Never go alone	41.4	74.5	21.2	43.4
Did not answer	1.5	—	8.1	15.1
Desire to become familiar with the 'other side'[d]				
Want to become much more familiar	25.7	36.2	67.7	79.1
Want to become slightly more familiar	31.4	37.2	12.1	7.5
Current familiarity is sufficient	31.4	24.5	8.1	3.8
No desire to become more familiar	11.5	2.1	11.1	1.9
Did not answer	—	—	1.0	7.5

Based on the following questions to the respondents:

Notes: a. 'When did you last visit the Arab (or Jewish) area of the city?'
   b. 'To what extent do you feel at ease when you visit the Arab (Jewish) area of the city?'
   c. 'When you visit the Arab (or Jewish) area of the city, with whom do you go?'
   d. 'To what extent are you interested in becoming more acquainted with the Arab (Jewish) area of the city?'

Given the Arabs' structural dependency on the Jewish sector, it is not surprising that even during periods of unrest Arab workers have no option but to traverse the ethnic frontier, whereas for the Jews visits to East Jerusalem are neither forced nor solitary. Still, the overall environment of Jerusalem is one of uncertainty, for interethnic violence may break out in unpredictable ways and this understandably intensifies fears for personal safety across the boundary. Jews are principally afraid of acts of Arab terrorism or the hostile demonstrations that periodically occur in East Jerusalem while Arab unease is intensified by frequent security checks and the possibility that Jewish reprisals may take place should a bomb explode during their temporary presence in West Jerusalem. Indeed, in the aftermath of the recent 'Arab uprising' (which has occurred since the research for this chapter was carried out), most Jews have demonstrated their lack of dependence on East Jerusalem by completely avoiding the Arab sector.

It is this geography of asymmetrical dependency that makes understandable the much higher degree of interest shown by Arab students in the Jewish side of the city than vice versa (see Table 10.4).[8] This is tellingly exposed in the simple fact that many more Arabs have learnt Hebrew than Jews have mastered Arabic. Generally speaking the differential interest in the other side further reflects the nature of majority–minority relations. Because the Arab minority *must* cross over to the Jewish majority sector, it has greater incentive than the Jews to become familiar with realities in the opposite side. The fact that Arab females, possessing a minority status within the minority group, expressed the greatest interest of all is in a way a further testimony to this general pattern.

Conclusion

In Jerusalem an 'ethnic boundary' continues to affect the everyday lives of Jews and Arabs alike. To be sure, it is not an impenetrable barrier; rather it bears spatial testimony to the nature of the social, political, and economic relationships that have emerged between the two communities. Doubtless there are other social divides in Jerusalem – say, religious–secular within the Jewish community, Moslem–Christian on the Arab side – but there can be no doubt that the most significant boundary remains that between the two rival national groups. The different cognitive maps of urban Jerusalem that each of these groups possesses is no less a reflection of social segregation as of the persisting political divisions – political divisions which condition the asymmetrical relationships that are themselves built upon Jewish control and Arab dependence.

Jews and Arabs have different mental maps of Jerusalem – a clear case of divided perception in a united city. This cognitive disjunction is a result of the complex interweaving of three basic strands. First, there is a 'segregation effect' whereby each group knows its own space best. Second, there is a 'majority–minority effect' whereby Arabs, having a greater dependency on the Jewish area than do Jews on the Arab, display the greater familiarity with the other territory. Finally, a 'group-performance effect' is demonstrated by a more developed Jewish ability to specify accurately various locations within the city (particularly within their own sector) – competences that may reflect different modes of training and education among the two ethnic groups. Indeed, the difficulty that some Arabs had in using maps to locate the particular sites specified in the questionnaire (and which meant that responses were restricted to verbal description) suggests that Arabs are less comfortable with the cartographic tools of locational representation.

No doubt the mental maps that Jews and Arabs have of Jerusalem have sources of compilation and modes of construction similar to those found in other ethnically divided cities throughout the world. Taken together these maps confirm the limited spatial familiarity of individuals, the linkage between urban mental images and location of residents and activity patterns, and, more particularly, the influence of ethnically related cultural behaviour. Classically, groups higher up on the socio-economic scale are believed to have a markedly greater degree of urban familiarity than those located lower down. The Jerusalem case, however, puts significant restrictions on this generalization: for, taken overall, Arab familiarity with the city is only marginally less than the Jewish – not quite what the conventional model would lead us to expect. Admittedly, in the divided city of Jerusalem, it is the higher socio-economic group – the Jews – which knows its *own* sector best, but this is counterbalanced to a significant degree by the fact that Arab students are more familiar with the Jewish sector than are the Jews with the Arab. The explanation for this deviation from the model is rooted in the importance of political factors as well as the dominant–subordinate relationships that form the everyday reality of Jerusalem.

In Jerusalem, it is plain, geographies of the mind have played no small part in shaping the social fabric of the city.

Notes

1 A useful overview of residents' 'urban perceptual maps' may be found in D. Ley, *A Social Geography of the City* (Harper & Row, New York, 1983). A more comprehensive discussion is to be found in P. Gould and R. White, *Mental Maps* (Penguin, Harmondsworth, 1974). Of the studies relating social and ethnic status to urban familiarity, particular note should be made of P. Orleans, 'Differential cognition of urban residents', in R. M. Downs and D. Stea (eds), *Image and Environment: Cognitive Mapping and Spatial Behavior* (Aldine, Chicago, 1973).

2 Using the distribution of Jews and non-Jews at the level of Sub-quarters in 1985, a dissimilarity index between the two groups of 95.68 has been calculated. Based on: *Jerusalem Statistical Abstracts* (1986).

3 Among the many works dealing with the urban and social development of Jerusalem in general and Jewish–Arab relationships in particular, see D. H. K. Amiran, A. Shachar, and I. Kimhi (eds), *Urban Geography of Jerusalem* (Walter de Gruyter, Berlin, 1973).

4 This survey was carried out as part of a wider project concerning interrelationships between the Jewish and Arab sectors in Jerusalem. For a review of the findings of this study, see M. Romann, *Interaction between the Jewish and Arab Sectors in Jerusalem: Economic and Spatial Aspects* (Jerusalem Institute for Israel Studies, Jerusalem, 1985) [in Hebrew]. The survey covered a sample of pre-final year students in six schools in the Jewish sector and five schools in the Arab. In total, 317 pupils were surveyed, 164 Jewish and 153 Arab. The choice of schools was carried out in such a way that the sample would include normal high school students, those at technical schools, and others of different educational standards, as well as schools with different locational proximities to the 'other side'.

5 See S. Shai, 'Interactions between the Jewish and Arab sectors in Jerusalem: multi-factor analysis of striking roots in the city', Israel Institute for Applied Research Internal Report (1983) [in Hebrew].

6 The weighting procedure was as follows: the Jewish familiarity percentage for sites in the Arab area (20.1 per cent) is divided by the Jewish familiarity percentage for their own sector (83.9 per cent) to yield a value of 0.2395. Likewise, the Arab familiarity percentage for sites in the Jewish area (26.0 per cent) is divided by the Arab familiarity percentage for their own sector (64.9 per cent) to give a value of 0.4006. These two values are then multiplied by 100 to give 24 per cent and 40 per cent respectively – a difference of 16 per cent.

7 It might be said that asking respondents to locate sites on a map would overcome the problem of vagueness. However, this option was excluded because it was found that many students, particularly Arabs, were unable to use this method to good effect.

8 Selective mutual awareness in bordering areas, like Abu Tor, is discussed in A. Weingrod, 'An anthropological study of Arab–Jewish relationships in Jerusalem', unpublished Final Report to the Ford Foundation, 1985.

Part four

Re-evaluation

Chapter eleven

Thoughts, words, and 'creative locational acts'

Chris Philo

> In so important a matter as the selection of a site for a county asylum the reports of county surveyors and the judgement of county magistrates is not always to be depended upon. The law places the ultimate responsibility in the hands of the Commissioners [in Lunacy], they have sometimes decided upon reports which have not proved trustworthy.[1]

In recent years I have been researching the historical geography of what I call the English and Welsh 'mad-business', and in so doing I have examined the spatial distributions exhibited by past establishments sheltering, restraining, and caring for the mentally distressed.[2] Some of these establishments have been specialist madhouses, lunatic hospitals, or asylums, whilst others have mixed up the mentally distressed with all manner of other sick and 'misfit' people; but almost all have possessed intriguing geographies with distinctive urban, rural, local, regional, and environmental affinities. It has rapidly become apparent to me, though, that these affinities cannot be 'explained' solely by abstract principles of spatial organization or by the workings of an underlying socio-economic formation, and I would argue instead that 'so important a matter' as locating a house for the mentally distressed must be seen as but one amongst many institutional practices – including plans and architectures – that are carefully thought through and decided upon by particular entrepreneurs, 'mad-doctors', administrators, and politicians.

In effect, then, the variety of locational analysis to which I subscribe requires a sensitivity to the tangled realm of thoughts, decisions, and behaviours, and this means that I am heartened by the behavioural 'turn' that various geographers have introduced into the curriculum of the location school over the last twenty or so years. My pleasure is not unqualified, however, since in practice I find many of this school's assumptions, claims, and models to be unconvincing and even positively unhelpful. In part this may be because the institutions that I study – given their multi-faceted social role in relation to the demanding phenomenon of 'madness' – are so different from the normal round of farms, settlements, shops, and factories that conventional behavioural location theories will never be of any great value to me, but I cannot help feeling that the reasons for this failure of the theo-retical machinery lie at least as much with its own cogs and springs as with the peculiarities of my chosen subject matter. As a result, the first part of this chapter involves a brief description of the behavioural 'turn' in location studies and then a selection of critical comments, most of which concentrate on how the *contents* of locational decisions tend – if they are not neglected altogether – to be trivialized and viewed in isolation from any broader con-ceptual or material context.

My objective here is not simply negative, though, for I hope to introduce some fresh theoretical materials that will counter these tendencies to trivialize and isolate the locational decision. One promising avenue of enquiry, I will suggest, is to follow Michel Foucault's focus on the spoken and written *discourses* that have nearly always been implicated in past institutional and other societal developments,[3] and in the third part of this chapter I offer a detailed account of how these discourses can be seen to comprise an inherently serious conceptual context – itself anchored in a maze of different material contexts – out of which specific dis-courses and behaviours periodically emerge. Moreover, it has become increasingly obvious to me that the ideas of William Kirk can furnish a sort of bridge – a bridge that I investigate in the second part of the chapter – between behavioural location studies and a Foucauldian focus on discourse. Hence, whilst various passages from Kirk's seminal essays of 1952 and 1963[4] clearly anticipate the behavioural 'turn', they largely avoid the traps of trivialization and isolation by situating locational decisions against the 'behavioural environment' – itself anchored in the objects of a host 'phenomenal environment' – peculiar to a given people in a given time and place. And, in addition, Kirk's concern for the conceptual organization of the contents deposited in this 'behavioural environment' parallels Foucault's concern for the dispersion of discursive acts into individual discourses, whilst his methodologi-cal allegiance to a *holism* derived from Gestalt psychology paral-lels the *archaeological holism* that lies behind much of Foucault's earlier research.

The behavioural 'turn' in location studies

In the mind of management, the presence of skilled labour has perhaps been the outstanding advantage of a location in Northamptonshire and Leicestershire, and it is management which, intuitively or otherwise, makes locational decisions.[5]

For as long as geographers have been studying the spatial distributions traced out by farms, settlements, shops, factories, and other human productions, there has been some recognition of the human 'variable' – the realm of thoughts, decisions, and behaviours – that intervenes between the 'facts' of the external world and the locations which these human productions end up occupying. In the above quotation, for instance, P. R. Mounfield identifies the role played by the 'mind of management' in shaping the geography of the East Midlands footwear industry, and such a quotation could be multiplied many times over from the literatures of economic and historical geography. On the other hand, it might be argued that the onset of geography as 'spatial science' during the 1950s and 1960s led to the downplaying of thoughts, decisions, and behaviours because researchers supposed human productions to be spatially distributed according to universal principles of spatial organization. Human decision-makers responsible for locating farms, settlements, and the like were hence treated as 'automata' responding to the 'economic-geographic' logics spelled out by the models of Christaller, Von Thünen, Weber, and a host of other spatial scientists. Seen in this light, any reference to the 'mind of management' could be constructed as no more than a decorative irrelevancy.

This blindness to human involvement in locational acts did not pass unchallenged, however, for it prompted a number of geographers to dispute the practice of a narrowly conceived locational analysis. In particular, it led Allan Pred to lay the foundations for a 'geographic and dynamic location theory' sensitive to the way in which 'any economic-geographic distribution, any array of land uses, any pattern of spatial interaction, is an aggregate manifestation of individual decisional acts made at the personal, group and/ or firm level'.[6] Using the well-known device of the 'behavioural matrix', Pred emphasizes that most spatial distributions, arrays, and patterns reflect the activity of decision-makers who operate not as 'rational economic men', but as agents with imperfect knowledge, faulty reasoning powers, and a tendency to be 'satisficers' rather than 'optimizers'. A number of objections can be raised to these formulations, principally because they treat actual chosen locations as mere 'deviations' from ones that theory would judge to be economically optimal; but Pred's achievement is still to offer an account of 'behaviour and location' that goes some way beyond the 'geometric determinism' of spatial science.

The consequence, I would argue, was to provoke a behavioural 'turn' in the field of location studies that has now filtered down into the discipline's more systematic branches, whether concerned with agriculture, retailing, transport, recreation, welfare services, or whatever. It is probably in the texts of industrial geography that a behavioural approach has come most markedly to the fore, with David Keeble suggesting in 1977 that behavioural studies had become 'arguably the most important single area of industrial location research'.[7] Writing some ten years earlier, Ian Hamilton had hinted at the possible scope and richness of adopting a behavioural approach sensitive to human involvement in industrial location, and in so doing had highlighted the need to avoid seeing locational acts solely in economic terms:

> it cannot be denied that the economics of procurement, production and distribution combined is *the* important – even deciding – location consideration. To infer that it is the only one, however, and accordingly to explain location mechanically as many location models do, is to deny that man is human. . . . Man's thinking on the location of industries is formed under pressure from a vast array of human practices, prejudices, habits, laws and systems.[8]

But it does not strike me that the insights and 'promise' of this passage have ever been successfully translated into concrete research, and I am unhappy that most behavioural studies of industrial location have tended either to neglect or to trivialize and isolate the *contents* of particular locational decisions.

Neglecting, trivializing, and isolating the contents of industrial-location decisions

Given the apparent centrality of the locational decision to behavioural studies of industrial location, it is strange to find how rarely the actual contents of particular decisions (the factors taken into consideration, their interpretations, the lines of reasoning pursued around them) are dealt with at any length. This neglect may be partly symptomatic of the disfavour that has now befallen the older 'descriptive behaviouralism' found in papers such as Mounfield's, given that the empirical findings of these papers seemingly 'contribute nothing to generalization or theory, belonging to the old case-study tradition of geographical inquiry'.[9]

Hence, rather than take Pred's intervention as a warrant for – as it were – returning to the details of the East Midlands footwear industry, many researchers who subscribed to the project of effecting a behavioural 'turn' in industrial geography shifted instead to the objective of forging an all-inclusive 'behavioural theory of the firm'.[10]

As a result much energy has been expended on trying to enumerate all of those various factors – all of those variables to do with firm 'organisation, ownership, size, decision-making procedures and so on'[11] – that supposedly influence the behaviour of the industrial enterprise, and in the process Hamilton's concern for actual 'thinking on the location of industries' has been largely submerged beneath a welter of boxes-and-arrows diagrams culled from the literature of organization theory.[12] Indeed, the studies involved here have spawned countless models that, as Keeble complains, tend to be 'not much more than classificatory descriptions of the sequences through which firms may (or may not) pass in making some sort of locational decision'.[13] A pioneer in this area of enquiry was Robert McNee, who long ago suggested that the spatial impress of a large industrial corporation must reflect its 'functional organisation',[14] although it might also be noted that McNee supposed his perspective on industrial geography to have a 'humanistic' flavour.[15] Nonetheless, McNee has often discussed the 'organised systems' of corporations in terms that owe much to the systems thinking of Talcott Parsons and A. Rapoport,[16] and it is this leaning towards 'functionalism' that has been imitated in numerous other studies concerned with enterprise organization and behaviour. This manœuvre has been commented upon by various authors,[17] as too have the serious theoretical problems associated with functionalism,[18] and all I wish to reference here is Neill Marshall's objection that 'functionalist work glosses over the relativistic aspect of reality, neglecting the indexicality of the social situation, that is, its dependence on who said what, to whom, and in what context'.[19] And it is surely inevitable that a lack of sensitivity towards 'who said what, to whom and in what context' will go hand-in-glove with a neglect of the contents written into particular locational decisions.

Having said this, it is still possible to find a few studies that do lend some attention to these contents, and in this respect it is important to mention the lengthy lists of factors that enterprise managers state – when being interviewed or when completing questionnaires – to be an influence on their locational decisions, past, present, and future. The sorts of factors cited obviously vary greatly depending on the type, scale, and ambitions of the enterprises under scrutiny, but perhaps the most trumpeted finding – certainly where smaller businesses are concerned – is the apparent importance of non-economic considerations such as the presence in a locality of helpful personal contacts, plentiful amenities, and a pleasant overall environment. Researchers such as Howard Stafford[20] and David Keeble have been especially vocal in pushing this claim to the fore, and it is instructive to examine the latter's attempt to 'prove' that recent changes in the manufacturing geography of the United Kingdom have been causally related to what he calls the 'residential space preferences' (RSP) associated with different parts of the country.[21] The suggestion is that certain environments – as evaluated in terms of climate, scenery, and amenities – are attractive to industrial managers and prospective skilled employees alike, and that they thereby exert a significant locational 'pull' on many 'footloose' industries.[22]

The question must arise, however, as to whether or not these sorts of account are effectively *trivializing* the issues really implicated in most industrial location decisions. This suspicion appears to animate Anthony Hoare's worry that the interpretation of Keeble's 'much-stressed' RSP finding

> will depend on our opinion of the use of Gould and White's famous school-leaver perception surface of residential attractiveness to model entrepreneurial locational preferences [which is how Keeble arrives at countrywide measures of RSP]. . . . Do the views of adolescents with (presumably) little or no business acumen, asked to make very speedy decisions about an unforeseen and purely hypothetical problem of unconstrained locational choice, bear any relationship to the actual decision that the constrained, experienced [entrepreneur] will produce after much more thought?[23]

I think that Hoare has here identified a problem widespread in the behavioural location literature, in that there is often a mismatch between the locational acts under study – which are for the most part non-routine and non-everyday occurrences[24] – and the sorts of routine, everyday, and even reflex mental exercises, such as making a 'speedy decision' about a favoured environment, that researchers reckon to be informing these locational acts. Nevertheless, it is repeatedly avowed that trivial personal and social considerations comprise a virtual 'hidden agenda' that many managers are reluctant to admit as an influence on their supposedly 'rational' economic decisions;[25] but could it not equally be argued that, although the final selection of a site *may* reflect certain personal and social preferences held by an enterprise's managers, this final selection will usually be made from a prior list of several locations deemed feasible on more strictly economic grounds? As R. C. Estall and R. Ogilvie Buchanan observed some years ago:

> an entrepreneur who lets his business judgement be swayed by the sort of golf he wants to play or the sorts of social facilities his wife demands is backing himself against all the odds to be lucky enough to hit on a suitable location without having [first] considered its business possibilities.[26]

And, as W. S. Thatcher remarked even more years ago, 'many factories have been placed after careful thought and calculation, the Ford for instance: one cannot imagine golf influencing the Ford executive'.[27]

The question as to whether or not behavioural location studies trivialize the contents of locational decisions cannot be answered definitively, of course; but what I can suggest from my own research is that the geographies of institutions such as madhouses, lunatic hospitals, asylums, and modern mental-health facilities reflect decisions that – almost without exception – embrace not the trivial whims of a few managers, but much hard, serious thinking on matters of great social moment. And it therefore occurs to me that *all* behavioural location studies risk trivializing locational decisions precisely because they treat these decisions as strangely *isolated* from any other thoughts, ideas, opinions, or speculations that decision-makers might possess. It might be objected that debates over the siting of houses for the mentally distressed are necessarily implicated in a larger and richer conceptual context than is the case for most industrial enterprises. My view here, though, is that even behavioural studies of industrial location should consider the ways in which managers – and, for that matter, employees – perceive, discuss, and act upon such advanced capitalist phenomena as fierce industrial competition and the relentless quest for profit, the skewering of society into social classes that repel each other mentally and physically, and the growth of political mechanisms designed to manage the volatility of both economy and social structure. I should of course acknowledge that an awareness of such questions *is* beginning to touch those economic geographers sensitive to the twin realities of agency and structure,[28] and one specific example of this growing awareness is to be found in Lois Labrianidis's investigation of the 'tactics' – and in particular the locational tactics – conceived and operationalized by managers of the Greek tobacco industry when seeking to defuse worker dislike for new industrial processes and patterns of production.[29] In short, therefore, this sort of 'radical' extension of the behavioural 'turn' in location studies 'does not lead back to behavioural analysis of industrial systems, but resolutely forward to historical approaches in which behaviour itself (within the dialectic of agency and structure) becomes a subjacent element of the whole process of historical eventuation.'[30]

'A gigantic decision-taking model of a special geographical type'

> Behind many of the distributions we map and study – villages, farms, fields, crops, mines, routes, artefacts – lie countless locational decisions, some well-documented, others less so, but unless we are able to recreate the decision-taking situations which produced them the determinants of their spatial patterns will never be fully known.[31]

Although it is rarely commented upon, a keynote feature of Kirk's thinking over the years has undoubtedly been his concern for how geographers – and particularly historical geographers – tackle the subject of 'creative locational acts'.[32] In the 1952 paper he refers on several occasions to the processes involved in the siting of cities – whether these be the two ancient Greek city-colonies of Byzantium and Chalcedon, the Buddhist cities of south-east Asia, or the city of Singapore as founded by Stamford Raffles[33] – and in a much later paper he continues this theme by recounting the story behind the selection of the site for the Burmese capital city of Mandalay.[34] As he explains here, despite being founded in the late 1850s, Mandalay was located according to mysterious and time-honoured cosmo-magical techniques through which it was

> 'discovered' that during his wanderings before attaining enlightenment the Buddha had visited Mandalay Hill, a holy mount, and had prophesied that in the 2400th year of his era – which happened to be A.D. 1857 – a great Buddhist metropolis would be built at its foot. All other omens pointed to the same site.[35]

Leading from this 'story' Kirk moves to the simple point that Mandalay's 'mode of location' is one that cannot be modelled using the Christallan tools of the Western spatial scientist,[36] and in so doing he places a large question mark beside the universalism – the desire to discover and articulate all-pervading laws of spatial organization – that is central to the spatial scientific endeavour.

Furthermore, it is precisely this sort of realization that brings him to ask the following question:

How often, for example, in studying the location of cities do we use the word location in a static (*Yin*) sense, as an established place, an object of human action, a stage for human drama, rather than as an act of creation worthy of study in its own right, a 'locating' (*Yang*) phenomenon, a decision to locate, in which various potentials are built into the creative behaviour from the moment of conception, similar to a dive from a high board in which the entire history of the dive is built into the mind and body of the diver at the moment of take-off?[37]

He would thus seem to be calling for a fresh species of location theory cognizant of how the whole creative locational process is nearly always anchored in an originating vision – a vision that may boast grand, even cosmological proportions – where the societal function and future history of the thing being located receive long, hard, and profound attention.

These statements on locational issues should not be viewed in isolation from Kirk's better-known attempts to introduce the concept of the so-called 'behavioural environment' into the orbit of geographical enquiry. As is now familiar, he distinguishes between a 'phenomenal environment' supporting the entire panorama of natural and humanly created things, and a 'behavioural environment' where the facts of this phenomenal world are organized into conceptual 'patterns and structures' peculiar to particular peoples in particular times and places.[38] He mobilizes the notion of the 'behavioural environment' principally to soften the outlines of environmental determinism in the study of human affairs,[39] but he also mobilizes it to clarify the perceptual-cultural basis of 'creative locational acts'. He thereby supposes the 'behavioural environment' to be the site where an originating vision is crystallized, ready for translation into a deliberate locational act, and this means that an appropriate question is something like: 'What was the behavioural environment of Stamford Raffles when he chose the site of Singapore and took such pleasure in the sight of the Union Jack moving above the mounds of "the old city"?'[40] An additional manœuvre in this respect can be identified as the thumbnail sketch that Kirk provides for a model where the 'geographical environment' – as a heady concoction of phenomenal and behavioural ingredients – is

regarded as a problem-producing situation in which human communities are perpetually confronted by problems concerning their particular environments and obliged to take decisions which have spatial [i.e. locational], environmental consequences. It is a gigantic decision-taking model of a special, geographical type.[41]

This is not the place to discuss the details of this grand but flexible vision,[42] but I do wish to offer a few further remarks relating to the intellectual ancestry of Kirk's proposals.

It is a commonplace to cite Kirk's debt to the German school of Gestalt psychology that emerged in reaction to stimulus–response behaviourism during the 1920s and 1930s,[43] but what is less commonly appreciated is the extent to which the details of his constructions echo Kurt Koffka's attempt to delimit 'behaviour and its field' in the latter's *Principles of Gestalt Psychology*.[44] Indeed, Koffka himself thinks in terms of the intersection between a 'geographical environment' containing physical and human objects – the environment that Kirk refers to as 'phenomenal' – and a 'behavioural environment' where physical and human objects are perceived, acquire a meaning, and then serve to regulate the actual behaviour of a given organism.[45] Moreover, Koffka also uses a diagram to demonstrate the relationship between the 'geographical' and 'behavioural environments' (see Figure 11.1) that Kirk redraws almost exactly in his papers of 1952 and 1963.[46]

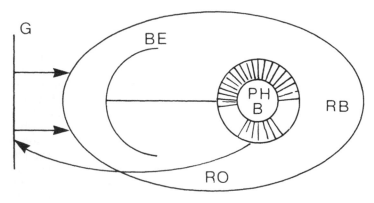

Figure 11.1 Kurt Koffka's 'Discoveries about behaviour and environment'

Note: 'G is the geographical environment. It produces BE, the behavioural environment; in this and regulated by it, RB, real behaviour, takes place, and parts of it are revealed in PH B, phenomenal behaviour. In some sense BE, RB and PH B occur within the real organism RO, but not in the phenomenal ego, which belongs with PH B. RO is directly affected by G and acts back upon it through RB'.

Source: K. Koffka, *Principles of Gestalt Psychology* (Kegan Paul, London, 1935), figure 2, p. 40.

Koffka's project boasts several features of interest to the social scientist, but for the present it will suffice to stress his distinction between 'molar behaviour' (or those numberless occurrences of what people conventionally term 'behaviour', like somebody crossing the road, going on a canal holiday, or revolutionizing the course of scientific investigation) and 'molecular behaviour' (or the physio-chemical operation of sensory surfaces, nerves, muscles, and glands that the biologist and behaviourist would term 'behaviour').[47] There is a propensity amongst conventional scientists to suppose that the former can somehow be reduced to or 'explained' by the latter, but Koffka objects that this line of reasoning is 'perfectly useless' for the purposes of the historian and – by extension – the geographer and the sociologist:

> Caesar's crossing the Rubicon: certain stimulus–response situations; Luther at Worms: so many others; Shakespeare writing *Hamlet*; Beethoven composing the Ninth Symphony; an Egyptian sculptor carving the bust of Nephretete, would all be reduced to the stimulus–response scheme. What then holds our interest in these occurrences? ... It is clear that such an aim (accounting for the specifics of these various acts) cannot be achieved if psychology begins and ends with molecular behaviour. Let us try molar behaviour instead.[48]

One implication of all this is that a psychology built around 'molar behaviour' would be entirely appropriate – at least in outline – for behavioural studies concerned with the specifics of 'creative locational acts', whereas a psychology preoccupied with 'molecular behaviour' would be entirely inappropriate. This is perhaps why much mainstream behavioural geography – with its behaviourist emphasis on the minutiae of stimulus–response equations filtering through 'black-box' minds – can say little about geographical actions other than routinized and trivial shopping trips,[49] and this is also why it is regrettable that most behavioural location studies still look far more to mainstream behavioural geography for guidance than to the 'molar' behavioural geography that Kirk effectively erects on the back of Koffka's ideas.

A second aspect of Gestalt psychology worthy of mention here is its methodological attunement to thinking in terms of identifiable 'organized wholes' – or *'gestalten'* – rather than in terms of phenomena that are chaotic, disjointed, or merely chained together in linear cause-and-effect relationships. As Koffka indicates,

> a gestalt is ... a product of organisation, organisation the process that leads to a gestalt. ... To say that a process, or the

product of a process, is a gestalt means that it cannot be explained by mere chaos, the mere blind combination of essentially unconnected causes.[50]

Kirk is obviously persuaded by these arguments, and adds his own voice to the Gestalt cause when suggesting that

> there is much to commend in the emphasis placed on the *principle of organisation* rather than *aggregation*. Here the debt to the Gestalt psychologists who insisted on the priority of the whole configuration over its constituent parts and stressed the relations between elements rather than the elements themselves ('the whole is greater than the sum of the parts') is evident.[51] (original emphases)

However, this *holism* is not solely a methodological device, since the Gestalt psychologists believe that the 'real' world beyond their offices and laboratories is itself constructed as an elaborate patchwork of abutting *gestalten*, and Kirk appears to subscribe to this ontological vision as well. In his eyes, therefore, the 'behavioural environment' peculiar to a given human community is not to be seen as a random arrangement of disparate notions stemming from many different 'authors', but as an edifice that organizes these many elements into some sort of discernible conceptual order; and neither is an individual location decision to be seen as a trivial, isolated event, but as a happening that must be understood in terms of the profundity and organization of the particular 'behavioural environment' from which it emerges.

Michel Foucault and the discursive contents of 'behavioural environments'

> Central to Foucault's work is a focus on discourse. From it flows his many insights, including ones derived from the relations between different discourses, between theory and practice, and between the desire for knowledge and 'profit' accruing to the ones who know. Foucault's search in the historical record avoids acceptance of taken-for-granted truths, thus discovering previously ignored or neglected beliefs and the practical consequences to which they lead.[52]

The suggestion must be that Kirk's various concepts – his 'phenomenal environment', 'behavioural environment', 'geographical decisions', and 'creative locational acts' – signpost a route beyond at least some of the deficiencies inherent in current behavioural location studies, but my impression is that further theoretical work is needed before these concepts can be mobilized successfully in concrete historical-geographical research. Many thorny questions remain as to the precise ontological status and characteristics of the 'behavioural environment',[53] for instance, and there is a grave danger here of foundering on the confused distinctions between individually and collectively, consciously and unconsciously held mental constructions.[54] Moreover, there is a danger that Kirk's methodological holism may repeat the functionalist failing of neglecting the specifics of particular situations when searching for overarching organizing principles,[55] and on a more practical level it is not immediately obvious from Kirk's writings just how a social scientist might set about identifying, investigating, and describing a given 'behavioural environment', its antecedents and its substantive effects.

In the pages that follow I aim to provide some guidance on these matters by discussing what might be termed the discursive contents of the 'behavioural environment', and in the process to engage with Foucault's attempts to reconstitute and critically evaluate past medical, psychiatric, penal, and sexual practices.[56] The point behind this engagement is that Foucault's struggle to render these practices intelligible depends greatly upon an 'archaeological' vision of *discourse* that is remarkable for opening up a whole new arena for the attention of the social scientist and the historian.[57] In brief, and as Colin Gordon explains,

> this project of an 'archaeology' is conceived as the study of forms of knowledge and rationality at the level of their material manifestation as bodies of discourse composed of finite sets of effective oral and written utterances. The aim is to render these discourses accessible to description and analysis as constituting a specific order of historical reality whose organisation is irreducible to either the history of the careers, thoughts and intentions of individual agents (the authors of utterances) or to a supra-individual teleology of discovery and intellectual evolution (the truth of utterances).[58]

In other words, Foucault supposes the field of discourse to be a hypothetical space occupied by speeches and sayings, academic papers and books, imaginative novels and poems, government publications, and all manner of other productions, the 'oral or written utterances' of which fall into distinctive though interwoven medical, legal, political, literary, and related 'bodies of discourse' – or individual discourses – homing in on important social matters such as 'madness' and its treatment.

Discourses: methodological and ontological considerations

Foucault first considers these discourses in a *methodological* light, and in so doing he acknowledges that the historian can only access past thoughts and actions through the discourses materially inscribed in a variety of historical *documents*. This means that 'the centrality of discourse in Foucault's method is less a philosophical principle than a practical requirement of historical research. ... No historian can avoid the primacy of the document. Foucault has given this fact a positive place in historical methodology.'[59] It is therefore surprising just how many historians *have* sought to avoid this 'primacy of the document', and have thereby latched on to one of two possible courses of action. The first of these stems from a positivistic belief in a neat one-to-one correspondence between the words of a document and some patch of reality lying beyond its pages, but this belief now appears untenable given Ferdinand de Saussure's demonstration that there is no simple linguistic connection between words and the concrete or conceptual things to which they ostensibly refer.[60] Further, it also appears untenable given the claims of a literary theorist like Roland Barthes, who identifies a modern 'crisis of language' whereby novels, poems, and other writings act not as representations of some external reality, but as little more than systems erected from signs that can only allude to or differ from one another.[61] Foucault evidently subscribes to this view, and it is revealing that Karel Williams draws upon Barthes, Foucault, and Saussure when warning historians about the error of treating documentary sources as unproblematic 'windows' on the past.[62]

The second course of action mentioned above entails psychologistic, psychoanalytic, phenomenological, hermeneutic, or even structuralist attempts to detect the true authorial intentions and desires, the deeper essences and meanings, or even the buried mental structures that are the real 'messages' hidden – as it were – between the lines of a document under scrutiny. However, just as Antoine Roquentin peered at the 'black, knotty mass of a chestnut tree root, realising that, faced with that big rugged paw, neither ignorance nor knowledge had any importance; the world of explanations and reasons is not that of existence',[63] so Foucault is 'haunted by the existence of discourses'[64] and is struck by the absurdity of positing any formal 'explanation' of discourse that remains external to its stubborn existence within the words themselves. This is not to suggest that he ignores the processes implicated in the production of discourse, since in practice he devotes much attention to the role of social situations, institutional sites, and authorial qualifications in this production; but what he does object to are those exercises for which a 'manifest discourse' is regarded as 'no more than the repressive presence of what it does not say'.[65] In other words, he objects to any exercise that posits and prioritizes an 'already-said' which precedes, shapes, but is not immediately present in the historical texts under study; and in place of such exercises he insists that

> we must be ready to receive every moment of discourse in its sudden irruption; in that punctuality in which it appears, and in that temporal dispersion that enables it to be repeated, known, forgotten, transformed, utterly erased and hidden far from all view in the dust of books. Discourse must not be referred to the distant presences of the origin, but treated as and when it occurs.[66]

Foucault therefore wishes to preserve the sovereignty of discourse in methodological terms, but he is also making a number of novel *ontological* claims that reflect his recognition of the world being *more* than 'purely and simply discourse'.[67] He thereby negotiates the sort of trap that ensnares an extreme, 'new critical' treatment of historical literary works, and which Charles Muscatine so ably describes in this passage:

> our problem is that the age of literary analysis – the New Criticism – seems to have reached a dead end. Conceived in reaction to a simplistic, 'positivistic' kind of [literary] history, it turned our attention to the text in and for itself and taught us to read poetry with the minute intensity that is now part of our standard equipment. The New Criticism taught us that the [conventional] parts of scholarship – editing texts, tracing sources – were ancillary to the great act of reading and elucidating the text in and for itself; that is the text's meaning was somehow hedged against historical relativism; that the literary work enjoyed a special *is*ness, a special ontological status. The new critical position, despite all it has taught us, can no longer be defended as an end in itself. ... Turning us away from a bad kind of history, it has tended to turn us away from history itself.[68]

The implication of this passage is not that researchers should forget the methodological insights of Saussure, Barthes, and the New Criticism, and neither is it that they should return to obliterating discursive specificities in the hunt for a great 'already-said': rather, it is that discourses should be seen as part and parcel of a much more extensive history in which peoples, economies, societies, politics, and culture have all played a role. Whilst the field of discourse *should* still be seen as bearing a 'special ontological status', it should also be understood to exist alongside all manner of other orders or levels of reality, and should thereby be regarded as an entity which is itself subject to influences emanating from other levels of reality and as an entity capable of exerting its own reciprocal influence – when individual discourses are heard, read, contemplated, and perhaps translated into practice by society's decision-makers – upon many other levels of reality.[69] Thus, an initial discourse concerned with the plight of the mentally distressed may be shaped in part by contemporaneous economic and social occurrences, whilst it may in turn lead to marked transformations in the institutional practices, plans, architectures, and geographies of the 'mad-business'.

The upshot of all this is to capture in a more profound fashion than heretofore the creative role of 'oral and written utterances' in the production of societies, and it is also to see words and documents not merely as 'windows' on history, but as an 'order of reality' inextricably bound up in the making and remaking of this history. It is true that a focus on discourse – on the tangibility of ideas made material in speeches and writing – provides a methodologically attractive way of accessing past (and even present) 'behavioural environments', but it is also the case that the field of discourse must be seen as comprising a vital ontological component of these given 'behavioural environments'.

The archaeology of discourse

It is instructive to turn now to the details of Foucault's treatment of discourse, since this will reveal both the extent to which he avoids the deficiencies of behavioural location studies and the extent to which his constructions parallel those of Kirk. The key text in this respect is *The Archaeology of Knowledge*, although it should be admitted at once that its pages are marked by what Foucault himself confesses to a 'somewhat bizarre machinery'.[70] It is important to be clear as to the precise status of this apparatus, however, for it is assuredly not intended to be a methodological blueprint dictating how all historians should unravel the discourses that feature in their substantive enquiries, and neither can it be taken – at least in any straightforward fashion – as an exposition of the 'archaeological' method informing Foucault's own historical investigations. Rather, *The Archaeology* is more a self-contained exploration into the ontology of discourse – into what discourses are, what their constituents might be, and how these constituents are then pieced together – and in consequence the study is not so much methodological as 'topical'.[71]

Foucault introduces the task of *The Archaeology* by suggesting that 'the question proper to such an analysis might be formulated in this way: what is this specific existence that emerges from what is said and nowhere else?'[72] Or, as he also asks, 'how is it that certain statements appear rather than others?'[73] The notion of the *statement* is pivotal to Foucault's project, given that he refers to it as the 'atom of discourse'[74] and devotes several pages to explaining why it differs from other, more formal 'discursive unities' such as the sentence, proposition, and speech act; and his crucial finding here is that

> one finds statements lacking in legitimate propositional struc-
> ture; one finds statements where one cannot recognise a
> sentence; one finds more statements than one can isolate
> speech acts. . . . In relation to all these descriptive approaches
> it plays the role of a residual element of a mere fact, of
> irrelevant raw material.[75]

This residuum overflowing from the dimensions of pure form is the content – the substantive message or 'things actually said' – that the statement hurries into the field of discourse, and this reveals that for Foucault the specific contents of 'who said what' in a given human situation are being brought very much to the fore.

There is a near infinity of 'things said' waiting in the archive for the historian's attention, of course, but Foucault narrows the number down by only considering those statements that are, by virtue of the context in which they are voiced, highly (self-) *serious* and deserving of study.[76] Hubert Dreyfus and Paul Rabinow acknowledge this tactic by calling Foucault's statements 'serious speech acts', and by indicating that

> any speech act can be serious if one sets up the necessary
> validation procedures, community of experts, and so on. For
> example, 'it is going to rain' is normally an everyday speech act
> with only local significance, but it can also be a serious speech
> act if uttered by a spokesman for the National Weather Service
> as a consequence of a general meteorological theory.[77]

And this means that Foucault is 'interested in just those types of speech acts which are divorced from the local situation of assertion and from the shared everyday background so as to constitute a relatively autonomous realm'.[78] As a result Foucault is deliber-ately leaving behind the routine, everyday home of myriad ordinary, trivial speech acts – although this is not to suggest that in particular situations such acts may not acquire great importance and 'serious-ness' for both their speakers and their listeners, and neither is it to specify some universal criteria for distinguishing the serious from the non-serious. Indeed, I would argue, in continuation of my earlier analysis, that this is precisely the sort of manœuvre that behavioural location studies will need to effect if they are ever to uncover the roots of most worldly locational decisions.

Foucault next supposes that these serious statements fall into a collection of overlapping but distinctive discourses, each of which is imagined to be a multi-dimensional edifice spanning a number of 'structural domains'. One of these embraces the objects spoken of, and is related to a non-discursive world that throws up matters of concern for serious discursive attention; one of which embraces the manner in which these objects are spoken of, and is related to the time, site, and author of the statements concerned; one embraces the sorts of concepts built into the discussion of objects, and is related to those essentially 'pre-conceptual' moments when authors 'choose' whether statements about objects are to be understood in terms of the succession of objects one after another, the coexistence of objects in a grand taxonomy, or whatever; and the last embraces the sorts of themes – or strategies – that colour the treatment of objects, and is related to overriding intellectual 'beliefs' such as the nineteenth-century philologists being convinced of a kinship between all the Indo-European languages.[79] In relation to this four-tiered edifice the main statement becomes 'not in itself a unit, but a function that cuts across a domain of structures and possible unities [i.e. the four tiers], and which reveals them, with concrete contents, in time and space'.[80] In describing the way in which statements snake across these four tiers – clustering both 'horizontally' and 'vertically' to produce identifiable discourses – so Foucault fills in the weird and wonderful details of his 'bizarre machinery' (see Figure 11.2, where I furnish my own visualization of this machinery and also introduce some additional Foucauldian terms).

How, in practice, however, does Foucault envisage statements spanning the levels of objects, enunciative modalities, concepts, and themes? – how does he know whether a given archive of statements forms part of a discourse, a whole discourse, or perhaps even more than one discourse?: in summary, what are the *rules* that he supposes to govern the ontology of the discursive field? Commentators of a positivist or functionalist persuasion will probably be disappointed by Foucault's answers here, principally because his solution is to equate 'rules of formation' with 'conditions of existence'. This means that, whilst many analysts strive to disinter the abstract *a priori* laws that supposedly determine the forms taken by actual sentences, propositions, and speech acts, the

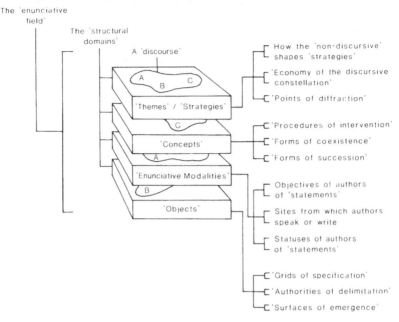

The 'enunciative field'

The 'structural domains'

A 'discourse'

A B C

'Themes' / 'Strategies'

C

'Concepts'

A

'Enunciative Modalities'

B

'Objects'

How the 'non-discursive' shapes 'strategies'

'Economy of the discursive constellation'

'Points of diffraction'

'Procedures of intervention'

'Forms of coexistence'

'Forms of succession'

Objectives of authors of 'statements'

Sites from which authors speak or write

Statuses of authors of 'statements'

'Grids of specification'

'Authorities of delimitation'

'Surfaces of emergence'

Figure 11.2 A visualization of Michel Foucault's 'bizarre machinery'

Note: A, B and C are serious 'statements'. They exist as 'atoms of discourse' in each 'structural domain' but the complete 'statement' – the 'enunciative function' – is comprised both by the 'atoms' and the 'vertical' connections between them. 'Statements' cluster both 'horizontally' and 'vertically' to give individual 'discourses'.

'archaeologist' claims only to describe the 'local, changing rules'[81] which legislate whether or not certain concrete statements belong to certain concrete discourses. Thus,

> [a] statement belongs to a discursive formation [or discourse] as a sentence belongs to a text and a proposition to a deductive whole. But, whereas the regularity of a sentence is defined by the laws of a language (*langue*), and that of a proposition by the laws of logic, the regularity of statements is defined by the discursive formation itself. The fact of its belonging to a discursive formation and the laws that govern it are one and the same thing ... [82]

Or, as Dreyfus and Rabinow observe, 'the rules governing the system of statements are nothing but the ways the statements are actually related'.[83] And yet Foucault does not envisage these rules tumbling preformed from the documents under study: rather, he supposes that an exacting examination of their pages is necessary before the researcher can begin to specify ways in which certain statements relate to certain other statements, identify similarities and dissimilarities in the objects spoken of, in the fashion of speaking, and in the concepts and themes used to organize the discussion, and suggest how statements of one kind (perhaps those tackling the location of asylums) connect up to statements of an apparently different kind (perhaps those tackling the nature of 'madness'). Furthermore, these rules emphasize not the essential sameness of 'things said', but the *differences* between them, and this is why Foucault supposes the only order underlying a discourse to be that given by what he calls the 'system of dispersion'.[84] The conclusion to be drawn is that Foucault's focus on the 'local, changing rules' governing the disposition of discourses within the 'behavioural environment' can provide a counter – a determinedly existential counter – to the 'big', *a priori* organizing principles latent in the treatment of the 'behavioural environment' by Gestalt psychologists.[85]

Another conclusion to be drawn is that Foucault's rules are greatly removed from a structuralist attempt to isolate both the 'basic elements' of some human activity and the laws by which these elements enter into manifest combinations. Dreyfus and Rabinow stress this point, and also offer a more fine-grained distinction between a *structuralist holism* inspired by Saussure's linguistic investigations and the *archaeological holism* that they detect in Foucault's projects:

> structuralist holism identifies elements in isolation and then asserts that the system determines which of the complete set of possible elements will be individuated as actual. In this case, one might say that the actual whole is less than the sum of its parts. Archaeological holism asserts that the whole determines what can count even as a possible element. The whole verbal context is more fundamental than its elements and thus is more than the sum of its parts. Indeed, there are no parts except within the field which identifies and individuates them.[86]

It would appear, then, that an 'archaeological' emphasis on the whole being more than the sum of its parts closely parallels the holism that Kirk draws from the Gestalt psychologists,[87] and this is perhaps a further legitimation for my proposal that discourses as described by Foucault comprise a vital component of – and are in a sense analytically compatible with – the 'behavioural environment' as described by Kirk.

Closing remarks and the case of the Cumberland and Westmorland public county lunatic asylum

> The decision of the Committee in selecting so healthy a site, on an elevated position, will make its wisdom felt throughout the whole future of the asylum.[88]

An act of 1845 (8 & 9 Vict. c.126) made it compulsory for every county and borough in England and Wales to make special provisions for their pauper insane charges, and by the mid-1850s most of the county and borough authorities had either complied with this requirement or were in the process of drawing up plans and purchasing land. In the second category here came the Magistrates of Cumberland, who had initially farmed their lunatics out to a private madhouse in the hope that this arrangement might 'relieve them from the duty of "building" a county asylum',[89] and also in this category came the Magistrates of Westmorland. By the early 1850s it had been decided that the two counties of Cumberland and Westmorland should unite for the purpose of erecting a public asylum, and in their annual report for 1853–4 the government's Commissioners in Lunacy noted that the Committee of Visitors to the proposed institution had purchased some 50 acres of land at Lowry Hill near Carlisle.[90] What the Commissioners did not report, however, was that impassioned objections to the purchased site were being voiced by a group of rated inhabitants living in and around this Cumbrian city.

These objections were written up in a memorial sent to the Commissioners in Lunacy,[91] but this document should not be dismissed simply as the 'bleating' of an interest group worried about possible increases in the 'county rate'. Indeed, the fact that one commentator writing in 1855 could describe financial support for the County Infirmary – a charitable institution located in Carlisle – as 'upon the most liberal scale'[92] might be mobilized against any portrayal of the memorialists as troubled solely by economic self-interest. Having said this, the memorialists did object to the asylum Committee paying £80 per acre for land valued by other 'competent authorities' at only £25 per acre, and they also objected to the Committee wasting £160 in a fruitless attempt at boring for water.[93] The substance of their complaints was not so obviously economic, though, since they referred to

> the bleak aspect and cold barren character of the soil, with an impermeable clayey subsoil, which no artificial drainage can remedy ... the close proximity of a dirty and noisome village ... the existence of brick-kilns on the south and east sides,

225

and the prospect of others being erected on the west and north, and immediately adjacent to the said proposed site ... the fact that the mines and minerals under the said proposed site belong to other parties, and have not been contracted to be purchased, and consequently the said site may be hereafter liable to the operation of these other parties.[94]

The key point to notice is that, whilst the possibility of disputes arising over underlying mines and minerals was largely a local problem, the references to both the site's physical geography and its relationship to industry and settlement were not composed in isolation, but in reflection of a much broader set of arguments about the environmental surroundings most conducive to the cure of mental distress. This state of affairs was even more apparent from a passage in which the memorialists announced their own vision of the features required by a good asylum site:

your memorialists are of the opinion that for the comfort and well-being of the poor unfortunate persons who are to become the inmates of the asylum, it is requisite above all things to have a plentiful supply of good water, a genial atmosphere and a salubrious locality, as very essential to physical health; and, if possible, where the landscape is varied and picturesque – the contemplation of which may be a source of daily pleasure to the unhappy lunatic, and may tend to awaken new emotions and create wider sympathies in his bosom, and thus materially conduce to his restoration.[95]

Lying behind and informing this particular passage was undoubtedly the whole web of thinking laid down in the proceedings of Select Committees, in the annual reports of the Commissioners in Lunacy, in the pages of the *Asylum Journal*, and in numerous specialist pamphlets and textbooks; and it is only by scouring these yellowing documents that the researcher can reconstruct something of the more extensive conceptual framework – the self-serious discourse which effectively organized many different discursive fragments into some semblance of order – out of which reflections upon the specific question of asylum siting were destined to emerge. Elsewhere I have endeavoured to spell out the details of the 'medico-moral' discourse that traversed the pages of these documents during the mid-nineteenth century, and which coloured contemporaneous exchanges over asylum siting,[96] and it is possible – though I do not have the space here – to indicate how this discourse was constituted from a concern for certain objects, from a certain way of speaking about these objects, and from a mobilization of certain pivotal concepts and themes.

What should also be noted in this connection is that the document prepared by the memorialists acted as more than simply a 'window' on the geography of the mid-nineteenth century 'mad-business', for it also exerted a significant influence upon the whole locational process.[97] The memorialists had claimed that several sites 'in convenient localities' throughout Cumberland could satisfy their own specifications, and in their annual report for 1854–5 the Commissioners in Lunacy stated that

> several other sites having been proposed on the part of the memorialists, it was suggested to the Visitors that any further proceedings towards the erection of the asylum should be suspended; a suggestion which was acquiesced in by the Visitors and the Justices in Quarter Sessions.[98]

Furthermore, an eminent civil engineer named Mr Rawlinson was instructed to inspect and evaluate the sixteen proposed sites, and he agreed that Lowry Hill was unsuitable but that several other sites – including one called the Garlands – were perfectly acceptable.[99] The final outcome was that the Lowry Hill lands were resold in order to finance the purchase of the Garlands, an estate located near a village called Carleton which was said to 'occupy a pleasant situation on the Penrith road three miles south-east of Carlisle',[100] and it was vaguely ironic that in his first report the medical superintendent to the institution (which was eventually opened at the Garlands in 1862) praised the Committee of Visitors for a site selection whose 'wisdom' would be 'felt throughout the whole future of the asylum'.[101]

What this example so neatly illustrates, then, is a chain of influence leading from an overarching 'medico-moral' discourse – which can be said to have possessed its own 'special *is*ness' – through the specific intervention and associated discursive fragment of the Cumberland memorialists, to the siting of an actual asylum on a particular patch of land blessed with particular locational attributes. A researcher could make little sense of what is involved here using the anonymous geometries of spatial science, and neither could he or she achieve much if supposing that site selection was a trivial, conceptually isolated matter shaped in large measure by the organizational machinery of asylum management. Rather, the researcher will only make progress by treating site selection as a 'creative locational act' spawned by a locational decision whose specific contents reflected – and were in a sense determined by – the serious, inclusive, and organized debates embodied in the words of oral and written discourses. Further, I would also conclude that there is no reason why this synthesis of ideas derived from both Kirk and Foucault should not have some theoretical purchase for geographical enquiries into the locations adopted by all manner of other human productions.

Acknowledgements

Various people have helped me in the preparation of this paper, but I particularly want to thank the editors, Felix Driver, Derek Gregory, Michelle Lowe, and Nigel Thrift.

Notes

1 J. C. Bucknill, 'Tenth report of the Commissioners in Lunacy to the Lord Chancellor', *Asylum Journal of Mental Science*, vol. 3 (1856), p. 20.

2 My Ph.D. thesis is provisionally entitled 'The space reserved for insanity: studies in the historical geography of the English and Welsh mad-business'. I use the term 'mad-business' to refer to the great diversity of institutions that have housed the mentally distressed at different times and in different places, and in so doing I follow the example of I. MacAlpine and R. A. Hunter, *George III and the Mad-Business* (Allen Lane, London, 1969).

3 Foucault, the deceased French post-structuralist thinker, has perhaps done more than anyone to translate themes from modern continental philosophy into the practice of historical enquiry.

4 The papers in question are W. Kirk, 'Historical geography and the concept of the behavioural environment', *Indian Geographical Journal*, Silver Jubilee Volume (1952), pp. 152–60; W. Kirk, 'Problems of geography', *Geography*, vol. 48 (1963), pp. 357–71.

5 P. R. Mounfield, 'The footwear industry of the East Midlands. The modern phase: Northamptonshire and Leicestershire since 1911', *East Midland Geographer*, vol. 4 (1967), p. 170.

6 A. Pred, 'Behaviour and location: foundations for a geographic and dynamic location theory', part I, Lund Studies in Geography, Series B: Human Geography, no. 27 (1967), Gleerup, Lund, p. 21.

7 See D. E. Keeble, 'Industrial geography', *Progress in Human Geography*, vol. 1 (1977), p. 306.

8 F. E. I. Hamilton, 'Models of industrial location', in R. J. Chorley and P. Haggett (eds), *Models in Geography* (Methuen, London, 1967), p. 368.

9 D. M. Smith, 'Decision-making', in R. J. Johnston (ed.), *The Dictionary of Human Geography*, 1st edn (Basil Blackwell, Oxford, 1981), p. 71.

10 This phrase – along with a criticism of 'descriptive behaviouralism' that parallels Smith's – is to be found in D. Massey, 'Towards a critique of industrial location theory', *Antipode*, vol. 5, no. 3 (1973), pp. 36–8.

11 See Keeble, 'Industrial geography', p. 305.

12 For a commentary and critique directed at this development see J. N. Marshall, 'Organisational theory and industrial location', *Environment and Planning A*, vol. 14 (1982), pp. 1667–83. One text particularly indebted to organization theory is P. McDermott and M. Taylor, *Industrial Organisation and Location* (Cambridge University Press, Cambridge, 1982).

13 D. E. Keeble, *Industrial Location and Planning in the United Kingdom* (Methuen, London, 1976), p. 4.

14 See R. B. McNee, 'Functional geography of the firm, with an illustrative case study from the petroleum industry', *Economic Geography*, vol. 34 (1958), pp. 321–37.

15 See R. B. McNee, 'Towards a more humanistic economic geography: the geography of enterprise', *Tijdschrift voor Economische en Sociale Geografie*, vol. 51 (1960), pp. 201–6.

16 See R. B. McNee, 'A systems approach to understanding the geographic behaviour of organisations, especially large corporations', in F. E. I. Hamilton (ed.), *Spatial Perspectives on Industrial Organisation and Decision-making* (John Wiley, London, 1974), pp. 47–57.

17 See, for instance, Marshall, 'Organisational theory and industrial location', pp. 1668–70.

18 See ibid. For a more general account and criticism of functionalism in geography see D. Gregory, 'Functionalism', in R. J. Johnston (ed.), *The Dictionary of Human Geography*, 2nd edn (Basil Blackwell, Oxford, 1986), pp. 165–7.

19 Marshall, 'Organisational theory and industrial location', p. 1671.

20 See H. A. Stafford, 'The anatomy of the location decision: content analysis of case studies', in Hamilton (ed.), *Spatial Perspectives on Industrial Organisation and Decision-making*, pp. 169–87, who emphasizes the 'personal contacts' factor and adds that, '[a]lthough this personal factor is always downgraded in normative economic models, it cannot be ignored when dealing with real world decision-making' (p. 178).

21 See Keeble, *Industrial Location and Planning*, esp. pp. 104–15, who employs a mathematical regression model to relate measures of recent changes in manufacturing employment and floorspace by 'subregion' to measures of RSP derived from the maps in P. R. Gould and R. R. White, 'The mental maps of British school-leavers', *Regional Studies*, vol. 2 (1968), pp. 161–82. Technically speaking this is not a behavioural study at all, but its underlying assumptions are clearly behaviourist.

22 See Keeble, *Industrial Location and Planning*, pp. 83–5.

23 A. Hoare, 'Review of Keeble's *Industrial Location and Planning*', *Progress in Human Geography*, vol. 1 (1977), p. 514. It should be acknowledged that Keeble is not entirely unaware of this problem with his analysis: see Keeble, *Industrial Location and Planning*, p. 104.

24 Industrial geographers regularly acknowledge that, except in the case of massive corporations, locational decisions are infrequent and are

not taken from 'within' a decision-maker's everyday work routine. A source of confusion here is perhaps the use of the term 'behaviour', which implies routine, everyday, little-reflected-upon activity, when the focus of concern is really 'actions', which imply rarer, less familiar, much-reflected-upon activity. This distinction is touched upon by I. G. Cullen, 'Human geography, regional science and the study of individual behaviour', *Environment and Planning A*, vol. 8 (1976), pp. 397–410.

25 The phrase 'hidden agenda' comes from Stafford, 'The anatomy of the location decision', p. 175. See also Keeble, *Industrial Location and Planning*, pp. 84–5, where he comments on the reluctance of key workers and managers to admit to such 'apparently unbusinesslike motives'.

26 R. C. Estall and R. O. Buchanan, *Industrial Activity and Economic Geography* (Hutchinson, London, 1966), p. 16.

27 See W. S. Thatcher, *Economic Geography* (English Universities Press, London, 1949), p. 142.

28 The pivotal paper outlining the need for geographers to take seriously *both* human agency and social structure when investigating all manner of human-geographical situations is D. Gregory, 'Human agency and human geography', *Transactions, Institute of British Geographers*, New Series, vol. 6 (1981), pp. 1–18.

29 See L. Labrianidis, 'Restructuring the Greek tobacco industry', *Antipode*, vol. 19 (1987), pp. 134–53.

30 M. Storper and A. J. Scott, 'Production, work, territory: contemporary realities and theoretical tasks', in A. J. Scott and M. Storper (eds), *Production, Work, Territory: The Geographical Anatomy of Industrial Capitalism* (Allen & Unwin, London, 1986), p. 12.

31 Kirk, 'Problems of geography', p. 370.

32 This phrase comes from W. Kirk, 'The road from Mandalay: towards a geographical philosophy', *Transactions, Institute of British Geographers*, New Series, vol. 3 (1978), p. 388.

33 See Kirk, 'Historical geography and the concept of the behavioural environment', esp. pp. 157 and 160. Note that it was an interest in 'creative locational acts' that led Kirk to consider the 'importance of perception problems in human spatial behaviour' when writing this early paper: see Kirk, 'The road from Mandalay', p. 388.

34 See Kirk, 'The road from Mandalay', esp. pp. 386–8.

35 Ibid., p. 386.

36 See ibid., pp. 387–8.

37 Ibid., p. 387.

38 See Kirk, 'Historical geography and the concept of the behavioural environment', pp. 158–60; Kirk, 'Problems of geography', pp. 364–9. Kirk's formulations have influenced – by their spirit, if not by their letter – much geographical work over the past 30 or so years, and this is even in such unexpected places as P. Newbury, *A Geography of Agriculture* (MacDonald & Evans, Plymouth, 1980), pp. 63–7.

39 Note, however, that Kirk wishes to retain a notion of 'environmental determinism', albeit one that allows for the refractions introduced by

Creative locational acts

what might be termed 'perceptual-cultural filters', in order, as he puts it, to counter the 'creeping paralysis of possibilism'.
40 Kirk, 'Historical geography and the concept of the behavioural environment', p. 160.
41 Kirk, 'Problems of geography', p. 368.
42 In outline Kirk considers the ways in which a human community perceives, makes decisions about, and acts upon its host 'phenomenal environment' – which he conceives of here chiefly in terms of population and resources – and he discusses changes in this environment which may or may not engender stresses in the community's 'behavioural environment', and changes which may or may not lead to the community taking ameliorative actions: see Kirk, 'Problems of geography', pp. 368–70 and figure 7, p. 370.
43 The history of the Gestalt school is briefly related in M. Henle, 'Wolfgang Köhler', in M. Henle (ed.), *The Selected Papers of Wolfgang Köhler* (Liveright, New York, 1971), pp. 3–10.
44 See K. Koffka, *Principles of Gestalt Psychology* (Kegan Paul, London, 1935).
45 See ibid., pp. 27–41.
46 Compare Koffka, *Principles of Gestalt Psychology*, figure 2, p. 40 with Kirk, 'Historical geography and the concept of the behavioural environment', figure 2, p. 159 and Kirk, 'Problems of geography', figure 5, p. 366.
47 See Koffka, *Principles of Gestalt Psychology*, pp. 25–6.
48 Ibid., pp. 26–7.
49 In a similar vein Bunting and Guelke worry that behavioural geography has lost touch with 'actual behaviour patterns', and that it ought to focus far more explicitly upon 'overt behaviour of geographic significance': see T. E. Bunting and L. Guelke, 'Behavioral and perception geography: a critical appraisal', *Annals of the Association of American Geographers*, vol. 69 (1979), esp. pp. 460–2. Note that this argument relates to my earlier remarks about a mismatch between non-routinized, non-everyday locational acts and the routinized, trivial mental exercises often reckoned to be their origin.
50 Koffka, *Principles of Gestalt Psychology*, pp. 682–3. See W. Köhler, *Gestalt Psychology* (Bell & Sons, London, 1930), esp. ch. VI on the 'Properties of organised wholes'.
51 Kirk, 'The road from Mandalay', p. 390.
52 E. Kurzweil, 'The neo-structuralism of Michel Foucault', in R. Wuthnow and others, *Cultural Analysis: The Work of Peter L. Berger, Mary Douglas, Michel Foucault and Jurgen Habermas* (Routledge & Kegan Paul, London, 1984), p. 141.
53 It must be remembered that the Gestalt psychologists who gave birth to the concept have tended to conduct formal analyses of how vision, memory, and learning are organized, and – notwithstanding their comments about 'molar behaviour' – they provide few clues about how the psychic materials that precede 'geographical decisions' might be organized.

231

54 I have already foundered on these rocks when elaborating upon the Jungian cast to Gunnar Olsson's thinking in C. Philo, 'Reflections on Gunnar Olsson's contribution to the discourse of contemporary human geography', *Environment and Planning D: Society and Space*, vol. 2 (1984), pp. 231–7.

55 In practice, however, Kirk's sensitivity to cultural variations appears to steer him clear of this trap.

56 Rather than provide a detailed bibliography of Foucault's historical studies I will refer the reader to the relevant chapters of A. Sheridan, *Michel Foucault: The Will to Truth* (Tavistock, London, 1980), which summarizes these studies using many of Foucault's own words and phrases. Central to my own research is M. Foucault, *Madness and Civilization: A History of Madness in the Age of Reason* (Tavistock, London, 1967).

57 Foucault offers his most sustained investigation of the field of discourse in M. Foucault, *The Archaeology of Knowledge* (Tavistock, London, 1972).

58 C. Gordon, 'Other inquisitions', *Ideology and Consciousness*, vol. 6 (1979), p. 34.

59 C. G. Lemert and G. Gillan, *Michel Foucault: Social Theory as Transgression* (Columbia University Press, New York, 1982), p. 38.

60 See F. de Saussure, *Course in General Linguistics* (Peter Owen, London, 1960).

61 See R. Barthes, *Writing Degree Zero* (Jonathan Cape, London, 1967).

62 See K. Williams, *From Pauperism to Poverty* (Routledge & Kegan Paul, London, 1981), esp. pp. 3–4 and pp. 18–19.

63 J.-P. Sartre, *Nausea* (New Directions Paperback, New York, 1964; Penguin, Harmondsworth, 1965; first published 1938), p. 185.

64 This is a phrase that Foucault used in an interview given in 1967: see Lemert and Gillan, *Michel Foucault: Social Theory as Transgression*, p. 21.

65 See Foucault, *The Archaeology of Knowledge*, p. 25.

66 Ibid. Another way in which he expresses this claim is to argue for documents to be treated, not as unproblematic traces of the past or as things to be interpreted or otherwise read behind, but as 'monuments' boasting their own internal organization, logic, and system (see esp. pp. 6–7).

67 See Lemert and Gillan, *Michel Foucault: Social Theory as Transgression*, p. 38.

68 C. Muscatine, *Poetry and Crisis in the Age of Chaucer* (University of Notre Dame Press, Notre Dame, Ind., 1972), p. 4.

69 This reading of Foucault obviously owes much to Gregory's ontological appropriation of 'theoretical realism': see D. Gregory, 'Suspended animation: the stasis of diffusion theory', in D. Gregory and J. Urry (eds), *Social Relations and Spatial Structures* (Macmillan, London, 1985), esp. pp. 327–30.

70 Foucault, *The Archaeology of Knowledge*, p. 135.

71 This point is stressed by Lemert and Gillan, *Michel Foucault: Social*

Theory as Transgression, who warn against receiving *The Archaeology* simply as a 'methodological programme', and add that, 'on closer reading, one realises that this purely methodological development is caught up in another theme. The book is also topical. Its subject is knowledge (*savoir*) and knowledge's relation to discourse' (pp. 48–9).

72 Foucault, *The Archaeology of Knowledge*, p. 28.

73 See ibid., p. 27.

74 See ibid., p. 80.

75 Ibid., p. 84.

76 This means that Foucault's statements are actually quite rare, and are thereby governed by a principle of 'rarity'. See M. Foucault, *The Archaeology of Knowledge*, esp. ch. III.4.

77 H. L. Dreyfus and P. Rabinow, *Michel Foucault: Beyond Structuralism and Hermeneutics* (Harvester, Brighton, 1982), p. 48.

78 Ibid., pp. 47–8.

79 See Foucault, *The Archaeology of Knowledge*, esp. chs. II.3–6 and chs. III.1–3. A commentary that I have found especially useful in this respect is M. Cousins and A. Hussain, *Michel Foucault* (Macmillan, Basingstoke, 1984), ch. 4, and see also Dreyfus and Rabinow, *Michel Foucault: Beyond Structuralism and Hermeneutics*, ch. 3.

80 Foucault, *The Archaeology of Knowledge*, p. 87.

81 The phrase 'local, changing rules' derives from Dreyfus and Rabinow, *Michel Foucault: Beyond Structuralism and Hermeneutics*, p. 55.

82 Foucault, *The Archaeology of Knowledge*, p. 116, and see also pp. 73–4.

83 Dreyfus and Rabinow, *Michel Foucault: Beyond Structuralism and Hermeneutics*, p. 55.

84 See Foucault, *The Archaeology of Knowledge*, esp. p. 37. Foucault is here displaying a thoroughly post-modern concern for *difference*, but he is also claiming that, just because things differ from one another rather than being manifestations of some deeper sameness, there is no reason why these differences should not be rigorously described and form the basis for further historical analysis.

85 Foucault's existential leaning has already been noted, and I am sure that he would follow Sartre in supposing 'existence to precede essence': see J.-P. Sartre, *Existentialism and Humanism* (Methuen, London, 1948), esp. p. 28. However, a Gestalt psychologist like Koffka argues that the 'essence' of a gestalt – by which he means its highly organized nature – 'is the reason of its existence', and the implication here is that there is some overarching law of organization that precedes and shapes the specifics of existence: see Koffka, *Principles of Gestalt Psychology*, p. 683.

86 Dreyfus and Rabinow, *Michel Foucault: Beyond Structuralism and Hermeneutics*, p. 55.

87 A key point is that for Foucault and perhaps also for Kirk the 'organized whole' is simply the particular human situation under study, and is *not* – as the Gestalt psychologists might claim – some

prior structure mysteriously 'hanging there' waiting to impose itself on the next human situation that comes along.

88 C. L. Robertson, 'Excerpts from asylum reports, 1863', *Journal of Mental Science*, vol. 9 (1864), p. 284.

89 See Visiting Justices of the County of Cumberland, 'Reports' (1846), reprinted as Appendix II in Anon. (probably C. L. Robertson), *Report of the Cumberland Lunatic Asylum at Dunston Lodge, Gateshead-on-Tyne* (Neill, Edinburgh, 1847), p. 27.

90 See Commissioners in Lunacy, 'Eighth annual report to the Lord Chancellor', *Parliamentary Papers*, vol. XXIX (1854), p. 15.

91 See Anon., 'Proposed county asylum for Northumberland and Cumberland', *Asylum Journal*, vol. 1 (1854), pp. 61–2. By some strange quirk this paper was mistitled, with 'Northumberland' being substituted for 'Westmorland'. Apparently a similar petition was prepared by fourteen of Carlisle's physicians, whilst the Board of Guardians was also reported to be very concerned (see p. 62).

92 See R. Asquith, *History of Carlisle Past and Present, and Guide to Strangers* (Thurnam, Carlisle, 1855), pp. 70–1.

93 See Anon., 'Proposed county asylum for Northumberland and Cumberland', p. 61. The failure to find water meant that the Committee was forced to recommend securing a supply from the Water Company at Carlisle some three miles distant (see p. 61).

94 Ibid.

95 Ibid., p. 62.

96 See C. Philo, '"Fit localities for an asylum": the historical geography of the nineteenth-century "mad-business" in England as viewed through the pages of the *Asylum Journal*', *Journal of Historical Geography*, vol. 13 (1987), pp. 398–415.

97 A recent history of the 'mad-business' that works with a Foucauldian notion of discourse similar to my own is M. Donnelly, *Managing the Mind: A Study of Medical Psychology in Early Nineteenth Century Britain* (Tavistock, London, 1983), esp. pp. xi–xii.

98 Commisioners in Lunacy, 'Ninth annual report to the Lord Chancellor', *Parliamentary Papers*, vol. XVII (1854–5), p. 540.

99 See ibid., pp. 540–1.

100 See T. Bulmer *et al.*, *History, Topography and Directory of Cumberland* (Snape, Preston, 1901), p. 284. Bulmer *et al.* provide a nice description of the Garlands Lunatic Asylum on p. 284.

101 See Robertson, 'Excerpts from asylum reports, 1863', p. 284. Note that the medical superintendent – a Dr W. P. Kirkman – reported on the lightness of the soil, remarking on its suitability for growing barley (see p. 284), but it might also be added that water supply remained a problem, and that an artesian well bored to a depth of 270 feet through the red sandstone was necessary to keep the asylum in fresh water: see Bulmer *et al.*, *History, Topography and Directory of Cumberland*, p. 284.

Chapter twelve

People and places in the behavioural environment

R. J. Johnston

In his seminal paper, 'Problems of geography', Bill Kirk introduced the concept of the behavioural environment to a wide audience.[1] In retrospect, this is sometimes interpreted as an early essay heralding the development of what are now known as behavioural geography and humanistic geography, and indeed there is much in what he said that is consistent with such a view. Nevertheless, the paper was not widely cited in the years after its publication.[2] The present chapter suggests why this was so, and extends Kirk's treatment in ways that are important for contemporary human geography.

The behavioural environment revisited

A principal goal of Kirk's paper was to find an answer to the question 'Are there ... problems of a specifically geographical kind, or for which geographers have a special responsibility?' (p. 364). His answer was that what identifies geography is its holistic approach to the study of environment, 'not as a thing apart but as a field of human action'. The separation of 'man' and environment in the systematic sciences creates an unreal dichotomy, therefore, and leads to simplistic theories such as those of environmental determinism and possibilism. Countering that separation required a focus on the 'geographical environment'.

Kirk divided the geographical environment into two components. The first – the 'phenomenal environment' – needed little discussion, since it was 'an expansion of the normal concept of environment to include not only natural phenomena but environments altered and in some cases almost entirely created by man'.[3] The second, and innovative, component was the 'behavioural environment', which is:

235

a psycho-physical field in which phenomenal facts are arranged into patterns or structures (*gestalten*) and acquire values in cultural contexts. It is the environment in which rational human behaviour begins and decisions are taken which may or may not be translated into overt action in the Phenomenal Environment.[4]

Like Lowenthal just before him, therefore (though he did not refer to that other seminal paper),[5] Kirk argued that the environment that stimulates human response is not necessarily constant, since it is not the phenomenal environment (the 'real world'?) that provides the stimulus, but rather 'the social and physical facts of the Phenomenal Environment ... [which] constitute parts of the Behavioural Environment of a decision-taker ... only after they have penetrated a highly selective cultural filter of values'.[6] Thus, 'facts' of the phenomenal environment and 'facts' of the behavioural environment need not be the same thing, so that the geographer who studies only the former may get a very wrong appreciation of the stimuli to which those being studied actually responded. As Kirk indicates with reference to the 'facts' of the existence of concealed coal measures in the British phenomenal environment, this only entered the behavioural environment when the measures were both explored and exploitable. (As Zimmerman had earlier stated, 'resources are not: they become'.)[7]

The contents of the geographical environment include many items that are created by human action, by decision-makers who act in the contexts of their behavioural environments. Hence, according to Kirk, the need to focus work on decision-taking, on how the world was

> perceived by human beings with motives, preferences, modes of thinking, and traditions drawn from their social, cultural context. The same empirical data may arrange itself into different patterns and have different meanings to people of different cultures, or at different stages in the history of a particular culture, just as a landscape may differ in the eyes of different observers.[8]

The behavioural environment expanded

It is easy to see why, in retrospect, Kirk's ideas are interpreted as early statements in the fields of behavioural and humanistic geography. His concept of a behavioural environment, and its importance to decision-takers, was central to the work of the school of hazard studies developed at Chicago by Gilbert White[9] – but they developed it independently of him. Why, then, did Kirk not have a greater explicit impact, especially during the 1970s?

The answer to that question lies, I believe, in the distance between Kirk's conception of geography as a discipline and that of the trend-setters in geography at the time. As already indicated, Kirk portrayed geography as a discipline concerned with the environment, and hence bridging the natural and social sciences. Most geographers who entered the profession in the 1960s and 1970s did not: they were either natural-science or social-science systematic specialists who, with the exception of those who studied the physical environment as a resource evaluated by human decision-takers, identified very few links (other than methodological) between the two types of science. Thus, the contents of Kirk's paper, if not the concepts, were not central to the concerns of the discipline at the time.

For the great majority of human geographers, then and now, the main contents of the behavioural environment are people and their artefacts. These were not specifically mentioned by Kirk, who illustrated the relevance of his ideas with reference to the 'geographical environment'. There is no reason to suppose that this means that Kirk did not believe that the concepts were more widely applicable (indeed, his collaboration with students of international relations – reported in 'The road from Mandalay' – clearly indicates the contrary). Rather, it reflects a definition of geography that was rejected by newer generations of scholars.

Much of the work by those newer generations has fallen into just the traps that Kirk identified for those working within his definition of geography. They defined a phenomenal environment, and assumed that it was the behavioural environment of the people that they were studying, rather than exploring the true nature of the latter. Today, that relatively simplistic error is rarely made. Nevertheless, the contents and construction of the behavioural environments have not been as fully explained as they ought. The remainder of this chapter sets out on the necessary journey of exploration.

People as the behavioural environment

Much contemporary human geography is concerned with interactions among people, within and between places. A great deal of it is concerned with depicting and seeking to appreciate the outcomes of those interactions – spatial patterns – rather than the interactions themselves, but interrelationships among people are fundamental to their study. Thus, people are major elements in the behavioural environment.

Probably the greatest volume of geographical research literature illustrating that statement relates to the study of urban residential patterns.[10] This shows that people and households defined in a variety of ways tend to live apart, with occupational, income, religious, age, ethnic, and other groups occupying separate, sometimes highly segregated, residential areas. The existence of the groups is largely taken for granted, so there is little exploration of the reasons for the separation, of why people want to live apart (or, perhaps more correctly in some situations, why they accept the operation of institutional mechanisms that produce the separation). A process of distancing has been promoted to account for the mechanisms of separation,[11] but this too is unsatisfactory in that it fails to address the key question fully. Two possible routes to a satisfactory answer are explored here.

Segmented selves and purified identities

Modern societies are extremely complex organizations. They are built on a very fine-grained division of labour and involve bringing together at high densities people from a range of cultural backgrounds. Those societies are interdependent wholes and yet, according to Tuan, people are unwilling to partake in the whole.

> As self-knowledge increases, so does a critical knowledge of nature and society, or the world. The world, subjected to critical evaluation, loses its objectivity and cohesiveness. An individual finds it more difficult to accept society's values and to partake in its affairs as a matter of course. . . . Given the freedom and the opportunity to explore self and world, few individuals in fact do so.[12]

Instead, Tuan argues, they retreat into segmented worlds to keep melancholy and boredom at bay. Aristocrats, for example, have gone in for conspicuous consumption, as with the building of palaces; the upper bourgeoisie have 'found salvation in home life and in the cultivation of intimate personal relationships'.[13]

But why? Tuan argues that fragmentation of wholes is a necessary consequence of analysing the world, that 'The human mind is disposed to segment reality.' And complex societies encourage it.

> A civilized society is large and complex. . . . Group cohesion and shared myths are tenuous, especially in times when external threats do not exist. A modern society's cohesion is almost constantly under the stress of questioning by its component

parts – institutions, local communities, and individuals. The material landscape itself provides suggestive cues. A primitive village gives the impression of being a single and rather simple entity if only because few manmade barriers such as curbs and walls are visible, and few places are reserved for exclusive functions. By contrast, in a large city the innumerable physical boundaries that keep people and activities in discrete areas forcefully remind us of the city's delimited and segmented character, its complex hierarchies of space.[14]

Thus, it seems, the only way that people can accommodate to large, dense, complex societies is by fragmenting them, and then withdrawing (in part, but probably not entirely) into one fragment. But on what criteria does the fragmentation occur?

Some insights to a possible answer for this question are provided in a similar analysis by Sennett, which is based in psychiatry.[15] (Tuan's work has no explicit disciplinary or philosophical foundation, though elsewhere he has espoused phenomenology, and his essays imply the existence of essences underpinning behaviour, mediated by cultural processes.)[16] During the period of adolescence, Sennett argues, individuals have to discover their own identities, to recognize, create, and sustain self-images. They do this, at least in part, by creating 'other images' – stereotyped representations of groups to which they do not belong. This is one element of a process of purification, of defining oneself by creating the same sort of segmented worlds to which Tuan referred.[17] By retreating into one segment, and avoiding contact with others, the individual

> has learned how to insulate himself in advance from experiences that might portend dislocation and disorder . . . the adolescent can sustain a purified picture of his own identity: it is coherent, it is orderly, it is consistent, because he has learned how to exclude disorder and painful disruption from conscious consideration.[18]

In other words, to live in a society we classify its members, associate with those in groups we identify with, and shirk contact with others. To the extent that the latter impinge upon our lives, we treat them as members of a group, all of whom have certain characteristics – those characteristics are social constructions, the myths that we generate about those who are not 'of us', and on which we base our behaviour towards 'them'.

Adolescence is, of course, a period of immaturity; it should be succeeded by maturity, when such simplifications are removed and people are able to accommodate the complexity and diversity (the disorder) of society. However, this does not occur in modern societies, according to Sennett, for they comprise not mature individuals in a wide variety of interactions but rather segmented, purified communities whereby people find self-identification through group identification:

> people draw a picture of who they are that binds them all together as one being, with a definite set of desires, dislikes and goals. The image of the community is purified of all that might convey a feeling of difference, let alone conflict, in who 'we' are.[19]

Societies thus become ossified into collections of such purified communities (which may or may not have a territorial base); instead of people letting a diversity of painful, confused, and contradictory experiences enter their lives,[20] communal solidarity is linked to patterns of avoidance. Order is imposed on society, at least in part, through the creation and maintenance of a particular spatial form.[21]

Such patterns of avoidance contain within themselves the seeds of major social conflict, as Sennett recognized. His book was written in the aftermath of the riots in US cities in the late 1960s, and was used to promote an anarchic approach to city planning that would dismantle the spatially segmented, purified communities, removing the system which ensures that 'although many people live together they seldom come into unknown, unplanned contact'.[22] Its achievement would be difficult, he recognizes, involving

> convincing men who have succeeded quite well in isolating themselves in warm and comforting shelters in the suburbs, or in ethnic, racial or class isolation, that these refuges are worth abandoning for the terrors of the struggle to survive together.[23]

It would involve putting people into areas where they felt different, where the complexity offered no easy sanctuary:

> Such a community would probably stimulate a young person, and yet scare him, make him want to hide, as it would everyone else, to find some nice, safe, untroubled place. But the very diversity of the neighbourhood has built into it the obligation of responsibility; there would be no way to avoid self-destruction in the community other than to deal with the people who live around the place. The feeling that 'I live here and I count in this community's life' would consist, not of a feeling of

companionship, but of a feeling that something must be done in common to make this conflict bearable, to survive together.[24]

The dissolution of the purified communities would thus both create tension within society and yet reduce conflict. It would create tension because it would insist on mutual accommodation, but would reduce conflict because the stereotyping of 'us' and 'them' would be removed: 'the experience of living with diverse groups has its power. The enemies lose their clear image, because every day one sees so many people who are alien but who are not all alien in the same way.'[25]

Introducing spatial scale

The discussion so far has been concerned with urban residential patterns, which are the products of social processes whereby people, individually and collectively, operate survival mechanisms for living in complex, high-density societies that involve spatial avoidance. The creation of 'us' and 'them' images within society leads to the creation of 'us' and 'them' places: in turn, this helps to sustain the images, since people can be characterized as much by where they live as by what they are, hence 'us' become 'people who live amongst us' and 'them' become 'people who live in other places'. The production and reproduction of segmented selves is aided by the creation of spatially defined purified communities, so that people are readily categorized in the behavioural environments of others according to their spatial origins, and readily categorize themselves because they are socialized to be 'one of us'.

The spatial origins discussed here so far are relatively small – portions of towns and cities. Are the ideas transferable to other spatial scales? One of the main ways of characterizing people is by their country of origin, and much social, economic, and, especially, political behaviour is built on such characterizations. People from different countries are clearly 'them', as opposed to 'us'. They are not, however, a 'them' that we have distanced ourselves from, since we were never anything but spatially separated. However, the facts of spatial separation stimulate the 'us' and 'them' characterization.

Both of these types of separation are contrived and involve people being raised and socialized in situations where the social stereotypes (with spatial connotations) already exist, and are sustained by reinforcing mechanisms such as informal socialization and formal schooling. Just as people are educated to accept a certain national identity, so they are brought up to accept a social identity which reflects both their social and their spatial background: segregation is a 'fact' to the child in the city, just as nationality is a 'fact' for the young citizen. The two types of separation differ as well. That tied to nationalism operates at a much wider spatial scale, and involves much more impervious boundaries than does urban residential segregation. Thus, although in each the majority of 'us' are ignorant about the great majority of 'them', so that we must rely on local stereotypes to define the latter, in the case of national separation it is difficult to overcome that ignorance through direct experience whereas with segregation within cities the few, fleeting, and impersonal contacts between 'us' and 'them' could be countered by various forms of 'social engineering'. (Direct experience could promote accommodation in the latter case, a different form of education than in the former, which must rely very largely on indirect experience – a powerful case for the teaching of geography.)

Attitudes towards 'them' at the national scale are based on stereotypes drawn either on very limited direct contact – usually with a few individuals only, who are undoubtedly not representative, if the concept of representative is valid in this context – or on communal images. The latter are created within our communities, and are transmitted to and by us as part of our culture. Some of those images are favourable – 'they' are like 'us', and are our friends/allies – whereas others are negative, presenting 'them' as threats. Our behaviour will be based on those culturally transmitted images, though as individuals we may not accept them in their entirety, or even in part; we may then seek to alter 'our' view of 'them', by our actions within our own culture.

The people who are in the category of 'them' are clearly part of our phenomenal environment. They are in our behavioural environment, too, although only indirectly. For most of the time they are irrelevant to the great majority of 'us', since they do not impinge directly on our daily lives. However, some people (politicians, bureaucrats, teachers, and so on) have to deal with 'them' on our behalf, because we live in an interdependent world. Our cultural image of 'them' influences how we direct our agents to act for us, in negotiations over the range of issues that bring different societies into contact (usually, though not invariably, through the institution of the state). Thus, international relations are conducted on the basis of behavioural images, the representation of populations via cultural stereotyping.

The creation of behavioural environments

What Sennett and, to a lesser extent, Tuan do is provide us with behavioural insights to the processes of distancing and stereotyping – insights that are crucial if we are ever to appreciate the sources of conflict in society, as a necessary first step to the building of a peaceful alternative. Such analyses are incomplete, however, because they fail to locate the origins of group differences in the structure of society. (Sennett's anarchic solution is also flawed because of the power gradient between such groups; in addition, as will be argued below, it is of course relevant to a particular spatial scale only.)

The structure of society and the construction of residential differentiation

In modern capitalist societies the main divisions are those of class, not just the simple bourgeoisie–proletariat division identified by Marx, but the many other (and often cross-cutting) cleavages that are linked to the distribution of economic power in advanced capitalist societies. These include occupation, income, education (level and type), housing tenure, and so on. All are not just functional divisions of society, but also bases (some, such as tenure, more important than others, such as education) for the segmentation of society, along the lines suggested by Sennett and Tuan. Thus, for example, members of different occupational groups create 'us' and 'them' images which lead to spatial distancing that parallels the social distancing implicit in the characterization of some occupations as superior to others. The same happens with tenure groupings – as superbly illustrated by the issue of the Cutteslowe Walls in Oxford, where residents of a middle-class housing estate constructed walls across roads to prevent the inhabitants of an adjacent council housing estate from walking through the former,[26] and by the many manœuvres undertaken by US suburban residents to exclude public housing from their neighbourhoods.[27]

According to the theses argued by Sennett and Tuan, such spatial distancing (and the conflicts that follow) would appear to be natural consequences of the socio-economic structuring of society. Divisions of society are created, necessarily; people characterize themselves as members of particular groups that are consequent on those decisions; and those groups create purified communities. The particular nature of the groups, and their origins in the relations of production, might seem to be irrelevant

243

to the process of purified community formation and distancing, but this is not so. Groups differ in their economic power, and those that have power are reluctant to yield it to others. Modern capitalist societies are (superficially, at least) meritocratic, in that the powerful positions (the 'best jobs') go to those best able to fill them. Education is therefore central to economic success, and so those who have succeeded seek to ensure that their children succeed too, by manipulating the educational system to their advantage. One way of doing this is to ensure that their children have privileged, if not monopoly, access to the best schools, and to the extent that schools draw on spatially defined catchments, so they seek to restrict access to such catchments for members of the out-groups – those who are of 'them' and not 'us'.

A further reason for distancing that is based on the economic divisions of society lies in the use of the dwelling as an investment as well as a 'machine for living' (i.e., in Harvey's terms, it has exchange as well as use value).[28] Part of the value of such investments lies in the quality of the properties themselves (their 'internal components') but much, too, depends on external factors – the nature of neighbouring properties, the quality of local schools, and so on. Thus, the membership of a segmented community is influential on the level – and, perhaps more importantly, the changing level – of its property values.

It would be taking the argument of the previous two paragraphs too far to imply that spatial distancing and the creation of residentially differentiated neighbourhoods is a *sine qua non* of capitalist cities. It is possible to envisage advanced capitalist cities without such spatial segmentation, but that is very unlikely to occur. This suggests that Sennett's and Tuan's ideas are irrelevant. But they are only partly so, however, because the local ideologies that they imply can be important elements in the creation, sustenance, and defence of purified communities; the creation of an 'us' and 'them' strategy, with negative stereotypes of 'them', can both aid the promotion of spatial distancing and provide a socially acceptable justification for it. In this way, the creation of purified communities is a strategy available to society rather than, as Tuan's phenomenology certainly implies and Sennett's psychiatry suggests, a natural outcome of social differences.[29]

Modern urban societies are not spatially segmented on economic variables alone, of course, as a generation of factorial ecologists has made very clear to us. A major criterion for the creation of segmented communities in many cities is ethnic status. With this, it might be argued, the purified communities are 'natural' rather than social creations, with 'us–them' strategies operating in both directions.[30] However, there is no doubt, too, that ethnic divisions are closely linked with socio-economic differentiation – no more so than in South Africa, where the system of apartheid has involved the reservation of certain (more 'desirable') occupations for 'whites only' and where black labour in particular is deliberately structured into a prototypical example of the 'industrial reserve army'. Again, therefore, although the ethnic differences and segmentation may be 'natural', there can be no doubt that they have been adopted and exaggerated to serve the purposes of the economically powerful; the racial superiority of whites and inferiority of blacks is central to the ideology of Afrikaner society and its use in promoting the political hegemony of white South Africans.

Other divisions within modern urban societies are just as potent, but they are not linked to purified communities and spatial segmentation in the same way. The best example of such divisions is that based on gender, which is used to promote the interests of males in a society that they have traditionally dominated. (Where gender and race coincide, acute problems often emerge, as illustrated by the position of black women in US society.) The role, and some would say necessity, of the nuclear family in society ensures that males and females are not permanently, spatially segregated, though of course they are (and certainly have been) for much of every 24 hours. (A major criticism of most urban social geography is that it is based on census data that refer to where people are in the middle of the night rather than during most of the hours of daylight!) At certain times of the day, the city is strongly segregated by gender.[31] So are many work-places, many homes, and many social gatherings. Again, is this 'natural', or is it part of a created world that promotes the hegemonic interests of the powerful (in this case males)?

The argument here, therefore, is that to understand how the mechanisms of segmentation and purified-community creation are operated, it is necessary to appreciate the societal context in which they occur. Capitalist societies are class-divided societies, and the perpetuation of the divisions (not only those of class but also the many others that are associated with it) is very substantially aided by the processes of distancing that Sennett and Tuan have identified. By associating themselves with particular territories in cities, and by excluding competitor ('undesirable'?) others from those territories, people seek to sustain their positions in society and to promote their interests over those of 'them'. As Sack has persuasively argued, territory can be used to divert attention from the reality of social relationships by expressing them as relationships between places;[32] Cox and McCarthy have illustrated how this takes place in the conflict over 'turfs', over the contents of neighbourhood purified communities.[33]

States, purified communities, and conflict

I have argued above that states, like urban residential neighbourhoods, are purified communities: they differ only in that the latter are created out of the complexity of cities whereas the states represent, in many cases, communities that already existed, independent of others. But why are states necessary, and why do they act as purified communities?

In the global world-economy that has evolved over the past four centuries, territorial boundaries that restrict the free flow of people, commodities, capital, and ideas are impediments to the processes of accumulation that form the heart of capitalism.[34] For certain periods, such barriers may be beneficial to some interests, since they can be used to guarantee monopolies over resources and, especially, markets, but such monopolies are ultimately constraining and spatial expansion is necessary. Thus, a system of states, of clearly defined territorial bodies, would appear to be counter-productive to the dynamics of capitalism.

To some extent this is so, but it is also true that states are necessary to capitalism. A strong argument supporting this claim is provided by Mann, who indicates that, without the existence of such an institution as the state, empirically independent of the spheres of both production and civil society, capitalism could not operate.[35] The regulation of many aspects of social relations and competition, without which the inherent anarchy of capitalism would lead rapidly to its own destruction, requires a state.[36] Furthermore, as Mann makes very explicit, the state is necessarily a territorial body: without clearly defined borders within which its sovereign powers are unchallenged, it could not undertake its necessary roles.

One of those necessary roles is the creation and maintenance of a social consensus within the state, for without the coherence that this brings the state could not successfully undertake its other major roles (securing the conditions for successful accumulation and legitimating the social relations on which that accumulation is based).[37] This involves the formulation and presentation of a state ideology, the creation of an 'us–them' situation which advances national (i.e. nation-state) interests as the interests of all of 'us', interests that can only be achieved through successful competition against some of 'them' and successful alliances with others. Harvey has provided a similar argument, based around the concept of regional class alliances, as necessary to the pursuit of capitalist goals.[38]

The state, then, is necessary to capitalism, as an institution that exercises economic power and, in order to do that, creates and sustains spatially defined ideologies. Its role thus involves the development and sustenance of a behavioural environment in which people are characterized as 'us' and 'them' according to whether they live within or outwith the state's boundaries, and behavioural responses appropriate to that categorization are encouraged.

The state is more than just a necessary institution in the economic organization of society, however, for a variety of reasons. One reflects its imposition on the cultural mosaic of the world. Prior to the development of capitalism as a global system, the world was divided into a very great number of separate cultural regions, each representing a particular local response to the people–nature and people–people interactions necessary to the construction of a society – a response that may well have been initiated in isolation from other societies but which would have been modified (probably substantially) as a result not only of internal events but also of contacts with other societies. Those societies were communities with a feeling of 'us-ness', a feeling that may have been promoted, especially in the largest, by state-like institutions.[39]

The importance of these communities is that some of them provide the foundations for modern nationalist movements, that seek to align the cultural map with the political map. Where a political map has been drawn with boundaries that are not coincident with those of the cultural map, the dominant interests within each state have sought to create a sense of national identity as part of their process of state-building, of creating social coherence within which capitalism can flourish.[40] Ranged against them, in some places, has been a second type of nationalism that seeks to redraw the political map – by creating a new state out of part of an existing one perhaps, by detaching part of one state and attaching it to another, or by amalgamating parts of several states into a new body. In such circumstances, two conceptions of 'us' come into conflict – that of the capitalist state and that of the nation which is not aligned with a state: the goal is to create nation-states, units of political organization spatially coincident with the purified communities recognized by some, if not all, of their inhabitants.

A further reason for needing to appreciate the role of the state lies in its autonomy. As Mann has argued, the fact that the state is a territorially defined institution gives it, or more accurately the politicians and/or bureaucracy that control it, a source of power that is independent of its economic role.[41] That power can be exercised in ways that are not directly linked to the economic role – although such exercise will be constrained by the need to carry out that role successfully. (State actions may stimulate crises of either accumulation or legitimation, or both, the resolution of which will require the curtailment of its autonomy.) Hence, bureaucracies may promote particular policies that they favour and which they can sustain because of their autonomous power-base. These may include foreign policies, based on particular attitudes to certain groups of 'them' and which lead perhaps to conflict: a necessary precursor to such policies must be the creation of the required attitudes towards 'them', through the promotion of a particular ideology.

Summary

Because of the particular definition of geography that he advanced, Kirk underplayed the importance of people in the behavioural environment, and especially the classification of people as either of 'us' or of 'them' that is characteristic of so much economic, social, and political behaviour. This chapter has taken Kirk's basic ideas and used them in the context of an (implicit) definition of human geography that places the study of people as central to its programme of work. In particular, it has suggested that appreciating how and why people are categorized within the behavioural environment, and how this both reflects and is reflected in the spatial structuring of that environment, is central to the pursuit of geographical understanding.

   Understanding spatial structures is a major task of human geographers. Central to that understanding, according to the writings of Sennett and Tuan reviewed here, is an appreciation of how people use strategies of spatial distancing to sustain images of themselves, in opposition to their images of others. Such images are collective, not individual – people promote views of them-selves as members of certain communities and, by definition, non-members of others. What Sennett and Tuan failed to provide, however, are accounts of how those collectives are formed – what criteria are used to define 'us-ness' and 'them-ness'. A preface to such an account has been attempted here. It is based on an understanding of the structure of capitalist society, but it has been stressed that, although that structure is the foundation on which communities are defined, it does not provide the only base. Hence, other divisions within society may be used to promote the concepts of 'us' and 'them'.

In many aspects of their daily lives, people deal with others (including very many others whom they never meet but whose existence and actions are relevant to their decision-making) not as known individuals but as representatives of one or more categories of individuals. Attitudes and behavioural patterns are attributed to them because those are the characteristics that make up the socially constructed images of those categories. We deal, then, not with individual people but with representatives of perceived ideal types. One of the bases for categorizing such ideal types is place of origin: people are assumed to have the characteristics associated with those coming from a particular place. Hence, activity in the behavioural environment involves dealing with people as representatives of places: we live in worlds comprising spatially segmented images.

Acknowledgements

I am grateful to David Sibley and Paul White for very useful criticisms and suggestions.

Notes

1 W. Kirk, 'Problems of geography', *Geography*, vol. 48 (1963), pp. 357–71. He first introduced the concept a decade earlier – in 'Historical geography and the concept of the behavioural environment', *Indian Geographical Journal*, Silver Jubilee Volume (1952), pp. 152–60 – a paper which he later wryly referred to as 'that often quoted but, I suspect, less frequently read, paper of mine': 'The road from Mandalay: towards a geographical philosophy', *Transactions, Institute of British Geographers*, New Series, vol. 3 (1978), p. 388.
2 Kirk, 'The road from Mandalay', reports that his ideas had an early impact on two US social scientists, having been conveyed to them by Oskar Spate.
3 Kirk, 'Problems of geography', p. 364.
4 Ibid., p. 366.
5 D. Lowenthal, 'Geography, experience and imagination: towards a geographical epistemology,' *Annals of the Association of American Geographers*, vol. 51 (1961), pp. 241–60. Lowenthal's paper was closely aligned to the study of perception by psychologists whereas Kirk saw his as contributing to a growing debate over the philosophy of geography rather than 'simply a restatement of the role of environmental perceptions or the definition of a research problem on

the frontier between human geography and psychology – which one could assume was the case from some later references to the paper': Kirk, 'The road from Mandalay', p. 389.

6　Kirk, 'Problems of geography', p. 366.

7　E. W. Zimmerman, *World Resources and Industries* (Harper & Row, New York, 1951).

8　Kirk, 'Problems of geography', p. 366.

9　See the recent reviews in R. W. Kates and I. Burton (eds), *Geography, Resources and Environment. Vol. II: Themes from the Work of Gilbert F. White* (University of Chicago Press, Chicago, 1986).

10　As summarized, over the years, in R. J. Johnston, *Urban Residential Patterns* (G. Bell & Sons, London, 1971); K. Bassett and J. R. Short, *Housing and Residential Structure* (Routledge & Kegan Paul, London, 1980); and P. L. Knox, *Urban Social Geography*, 2nd edn (Longman, London, 1987).

11　R. J. Johnston, *City and Society: An Outline for Urban Geography* (Hutchinson, London, 1984).

12　Yi-Fu Tuan, *Segmented Worlds and Self* (University of Minnesota Press, Minneapolis, 1982), pp. 196–7.

13　Ibid., p. 197.

14　Ibid., p. 7. See also C. Alexander, 'The city is not a tree', *Design*, no. 206 (1966), pp. 47–55.

15　R. Sennett, *The Uses of Disorder* (Penguin, Harmondsworth, 1973).

16　Yi-Fu Tuan, 'Geography, phenomenology, and the study of human nature', *Canadian Geographer*, vol. 15 (1971), pp. 181–92.

17　Sennett, *The Uses of Disorder*, p. 19 defines 'purification' as follows: 'a desire for a purification of the terms in which they [i.e. people] see themselves in relation to others. The enterprise involved is an attempt to build an image that coheres, is unified, and filters out threats in social experience . . . the degree to which people feel urged to keep articulating who they are, what they want, and what they feel is almost an index of their fear about their inability to survive in social experience with other men'.

18　Ibid., pp. 27–8.

19　Ibid., p. 38.

20　Ibid., p. 39.

21　D. Sibley, *Outsiders in Urban Societies* (Basil Blackwell, Oxford, 1981).

22　Sennett, *The Uses of Disorder*, p. 128.

23　Ibid., p. 146. Irwin Altman argues similarly with the statement that 'As the self-boundary is violated and as more interaction occurs than is desired under increased density, people attempt to reestablish boundaries. One way to do this is to withdraw from others': I. Altman, *The Environment and Social Behavior* (Brooks/Cole, Monterey, Calif., 1975), p. 184. The desired level of interaction is culturally defined, however, as pointed out in comparative studies of Hong Kong with other societies.

24 Altman, *The Environment and Social Behavior*, p. 178.
25 Sennett, *The Uses of Disorder*, p. 156.
26 P. Collison, *The Cutteslowe Walls* (Faber & Faber, London, 1963). See also B. T. Robson, 'The Bodley Barricade: social space and social conflict', in K. R. Cox and R. J. Johnston (eds), *Conflict, Politics and the Urban Scene* (Longman, London, 1982), pp. 45–61. In both of these cases, the 'excluders' are not necessarily among the most powerful in society: the processes operate at most levels, so that the relatively powerless distance themselves from the even less powerful.
27 These processes are discussed in detail in R. J. Johnston, *Residential Segregation, the State and Constitutional Conflict in American Urban Areas* (Academic Press, London, 1984).
28 D. Harvey, *Social Justice and the City* (Edward Arnold, London, 1973).
29 The issue of whether distancing is a natural, rather than a socially created, strategy is a very large one that has not been addressed here. The work of sociobiologists suggests that the processes of natural selection lead to certain traits becoming dominant – such as greed. Even if this is so, the existence of such generalized traits does not require particular outcomes (greed can be expressed in a great variety of ways), so that it is difficult to accept a case that distancing is 'necessary' – though it may be a very sensible strategy, as Sack suggests (see note 32): see also V. Reynolds and others (eds), *The Sociobiology of Ethnocentrism* (Croom Helm, London, 1987).
30 For a general discussion, see F. W. Boal, 'Ethnic residential segregation', in D. T. Herbert and R. J. Johnston (eds), *Social Areas in Cities*, vol. 1 (John Wiley, London, 1976), pp. 41–80; a detailed case study in D. Ley, *The Black Inner City as Frontier Outpost* (Association of American Geographers, Washington, DC, 1974). The categories 'ethnic' and 'racial' are frequently used as synonyms. Sandra Wallman – in 'The boundaries of race: processes of ethnicity in England', *Man*, New Series, vol. 13 (1978), pp. 200–17 – differentiates between the two, however: a boundary now called 'racial' is (or is perceived to be) fixed, immutable, hard; and a boundary called 'ethnic' must be (or is perceived to be) softer, more malleable, fuzzier by contrast (p. 215). Hence, it would seem, the harder definition by race can account for the persistence of black ghettoes, whereas white ethnic residential areas are less permanent: members of other races are always going to be part of 'them', whereas members of other ethnic groups within the same race can both aspire to and achieve membership of 'us'.
31 As discussed in Women and Geography Study Group, *Geography and Gender* (Hutchinson, London, 1984).
32 R. D. Sack, 'Human territoriality: a theory', *Annals of the Association of American Geographers*, vol. 73 (1983), pp. 55–74 and *Human Territoriality* (Cambridge University Press, Cambridge, 1986).
33 K. R. Cox and J. J. McCarthy, 'Neighbourhood activism as a politics of turf: a critical analysis', in Cox and Johnston (eds), *Conflict, Politics and the Urban Scene*, pp. 196–219.

34 This occurs at all spatial scales: see G. L. Clark, 'Law, the state, and the spatial integration of the United States', *Environment and Planning A*, vol. 13 (1981), pp. 1197–227.
35 M. Mann, 'The autonomous power of the state: its origins, mechanisms and results', *European Journal of Sociology*, vol. 25 (1984), pp. 185–213.
36 R. J. Johnston, 'Marxist political economy, the state and political geography', *Progress in Human Geography*, vol. 8 (1984), pp. 473–92.
37 G. L. Clark and M. J. Dear, *State Apparatus* (Allen & Unwin, Boston, 1984) provides a full discussion.
38 D. Harvey, 'The geopolitics of capitalism', in D. Gregory and J. Urry (eds), *Social Relations and Spatial Structures* (Macmillan, London, 1985), pp. 128–63.
39 R. J. Johnston, *Geography and the State* (Macmillan, London, 1982).
40 R. J. Johnston, D. B. Knight, and E. Kofman (eds), *Nationalism, Self-determination and the World Political Map* (Croom Helm, London, 1987).
41 Mann, 'The autonomous power of the state'.

Chapter thirteen

Mirrors, masks, and diverse milieux

Anne Buttimer

> Was Ihr den Geist der Zeiten heisst
> Das ist im Grund der Herren eigner Geist
> In dem die Zeiten sich bespiegeln . . .
> (Faust)

> (And what you call the Spirit of the Ages
> Is that the spirit of your learned sages
> The times a-mirroring . . .)

At Columbus, Ohio, where the Association of American Geographers met in 1965, there was obviously something new in the air. Eminent geographers and psychologists charmed a packed auditorium with ideas about environmental behaviour and perception. This new frontier was to welcome not only interdisciplinary research, but it was also to offer a common focus of curiosity to geographers of both 'man–land' and 'spatial' traditions. Why, even the age-old impasse between 'pure' and 'applied' orientations could be transcended.[1] Some 17 years later in San Antonio, Texas, the same Association hosted sessions on environmental perception. One caught a glimpse of the volume and variety of research which the intervening years had produced and, even more, one noted the drama of a selective migration of ideas back and forth across the Atlantic: Marxist, positivist, phenomenological, and structuralist approaches were juxtaposed, not always too harmoniously.[2]

The record of these 17 years should not, of course, be regarded as the only, or even necessarily the most typical expression of perception research in geography.[3] In European Schools the ebb and flow of interest has had a different rhythm, reflecting the vicissitudes of academic history and societal interest.[4] Some would ascribe these waves of interest to the charisma of key intellectual leaders or the diffusion streams of innovative ideas; others would

253

see them as responses to external challenge, mediated via the budget priorities of sponsors and audience. There is little doubt that the relative success and/or failure of twentieth-century attempts to probe the frontiers of human perception cannot be understood without reference to both internal and external circumstances.[5]

To many a researcher, the term 'perception', in operational terms, has connoted those images and 'subjective' preferences evident among some folk-out-there – Eskimo or Maori, migrant worker or college sophomore. An unspoken assumption is that the geographer's way of construing things is somehow more objective, detached, and potentially universal. Yet we all wear peculiar lenses on the world; we are all engaged in action, perception, and behaviour which reflect, and are reflected in, our particular historical settings. Why not then regard geographic thought and practice as a whole in terms of 'perceptions'?

This chapter examines the formation of geographers' own perceptions of reality. Data sources are the autobiographical accounts and published works of an international range of retired colleagues in various European and American countries.[6] The aim is not only to understand geographic thought and practice in contextual terms, but also to promote better self- and mutual understanding among practitioners in the field. The interpretative framework emerging from that study may indeed have heuristic value in assessing the significance of the environmental perception movement.

The rationale for focusing on researchers themselves rests on three main points. First, I suggest that perception research has been catalyst for a profound transformation in approaches to knowledge, from that of 'observer' to that of 'participant' in reality. Today, researchers face the inevitable challenge towards which the perception movement, among others, has pointed, viz. the hermeneutic circle and the need for scholars to become aware of their own *a priori* assumptions and methods. Second, I wish to explore the connections between disciplinary ideas and practices and the external contexts of societal interests towards which geographers have (explicitly or implicitly) addressed themselves. Three key themes – mirrors, masks, and milieux – will serve as an interpretative framework for this exploration. Third, I argue that perceptions of reality cannot be fully understood until they are set in the full context of a *genre de vie*. If this interpretative framework can facilitate the geographer's self-understanding, it is hoped that this might be catalyst for a better understanding of the *genres de vie* of folk-out-there. A few illustrative vignettes, drawn primarily from North America, will be cited as bases for some reflections about the harvest of perception research to date, and its prospects for the future.

Geography and life experience: a framework of interpretation

From observation to participation

Philosophers and historians of science have commented on the transformation of twentieth-century approaches to knowledge. From the quest for the foundations of knowledge and faith in the objectivity of scientific method, one has passed through a contextual phase, and is now in a 'post-foundational' era.[7] Echoes of this general transformation are discernible in the history of geographic thought and practice. In the early phases of discipline-making, geographers were mainly spectators, observers of pattern and process, and the key questions about knowledge were epistemological – namely, our modes of representing and elucidating phenomena. Already at century's turn, Heisenberg had noted that 'inevitable ripple' between observer and observed, and Niels Bohr had proclaimed his intriguing principle of complementarity in explanations of the physical world.[8] It was probably Thomas Kuhn's popularization of the paradigm idea that most effectively stirred awareness of context among social scientists and geographers.[9] And for empirical as well as logical example, what better models than Boulding's *The Image* and Lynch's *The Image of the City*?[10] The contextual turn brought about a blurring of those distinctions between subject and object, observer and observed, which were common belief in the era of observation. For geographers, the perception wave might thus be regarded as playing an intermediate role, ushering in an 'insider–outsider' phase during which energies were channelled towards elucidating differences in people's perspectives on nature, landscapes, and resources. Much of the analytical style and research models of this phase were still of the same vintage produced during the first (observation) phase. What marks the threshold between Phases Two and Three is the recognition of reflexivity, the acknowledgement of how social and other influences have been interwoven with the epistemological in the construction of conventional models.

The movement from observation to participation should not be understood in terms of linear progression or even chronological succession. One is speaking here of a conceptual transformation which found varied expression in different schools. All three phases are most probably evident in all schools simultaneously. Most practitioners would probably still position themselves most comfortably in Phase One, a small but energetic number might place themselves in Phase Two, and only a fragment would regard themselves as edging their way towards Phase Three. Several might move among these phases depending on the nature of their research curiosity. Philosophically speaking, however, it is perhaps helpful to identify the radical differences in dominant concern within these phases. Characteristically, the philosophical concerns of Phase One are *epistemological*, those of Phase Two *dialectical*, and those of Phase Three *hermeneutical*. Even a cursory glance at contemporary writings would show that all three are being simultaneously pursued today. There is the search for knowledge foundations and indubitable criteria of truthfulness,[11] the effort to demonstrate contrasts and conflicts of human interests,[12] and a growing awareness about the puzzle of diverse interpretations.[13] Inevitably the challenge of mutual understanding and communication among such diverse phases of effort grows.[14]

This three-phase schematization offers a broad diachronic frame on which to hang salient aspects of the perception movement. It demands a complementary – synchronic – approach which could shed light on relationships between perceiver and perceived at any particular moment. For the relationships between research (knowledge/truth) and life (being) are always mediated through the filters of academic lifeways.

Mirrors, masks, and milieux interests

At least three sets of curiosities arise when one considers the relationships between geographic perceptions and lived experience: (1) what kinds of truth (credibility) may one ascribe to research products, i.e. in what kind of 'mirror' is reality reflected for the geographer?; (2) in what kind of professional role and structure does the geographer construct his or her perceptions of reality, i.e. through what 'masks' does the geographer interact with the reality around?; and (3) what kinds of external or societal challenges and opportunities have played a role in shaping the nature of geographic research, i.e. to what 'milieux' interests does the geographer, explicitly or implicitly, address his or her efforts?

I use the term 'mirrors' here to denote key metaphors of truth which underlie geography's diverse models, theories, and paradigms of research. The basic approach has been inspired by Steven Pepper's theory of world hypotheses.[15] He claims that there are four distinct world-views which have stood the test of time in Western intellectual history: formism, organicism, mechanism, and contextualism. Each of these 'hypotheses' about the nature of world reality grounds its claims to truth, and its categories of analysis, on a root metaphor. In geography, I have suggested, one could find a reflection of these macro world-images in the root metaphors of *map*, *organism*, *mechanism*, and *arena* respectively.[16] This is not an exhaustive inventory of geographic 'mirrors' on reality, but these four metaphors serve to illustrate several key points: (1) the products of research conducted within the framework of any one of these metaphors cannot be submitted to judgement via the categories of another; (2) the appeal of a root metaphor does not rest exclusively on its cognitive claims – there are emotional and moral commitments which explain why researchers often cling to the same metaphor despite radical shifts of substantive focus or even ideology; and (3) a root metaphor can be regarded as source for a wide variety of theories, models, and paradigms; hence a look at differences of root metaphor may be far more revealing than a scrutiny of methodological or epistemological differences.

The expression 'masks' is used here in the sense intended by scholars of Symbolic Interactionist orientation,[17] and it refers to the roles played by practitioners in particular career settings, as teachers, writers, researchers, or consultants to planning. For some a career setting may best be elucidated by examining 'role' and its structural imperatives; for others, it may more appropriately be discussed in terms of vocational choice.[18] The advantage of the term 'masks' is that it helps to maintain the ambiguity or ambivalence of internalist and externalist perspectives on the social construction of knowledge.[19] Some would trace the authorship and practice of geography to the intentional choices of individuals; others would look at the archaeology of structures which produce and reproduce ideas and practices.[20] From the autobiographical reflections of senior colleagues, I have discerned four clusters of 'vocational meaning': *poesis* (critical, emancipatory writing and lecturing), *paideia* (teaching, textbook writing, leading field excursions, and so on), *logos* (promoting analytical rigour, law-seeking generalizations), and *ergon* (practical applications of geography to problem-solving, administration, and planning).[21] These categories of professional role, of course, face two ways – to the personal talents and dispositions of scholars on the one hand, and to the societally defined priorities of different periods on the other. Individuals have combined more than one of these, and have moved freely among them in the course of a career.

'Milieux' directs attention now to the concrete settings in which research is conducted, its sponsorship and audience, and the media through which its results are articulated. Emphasis rests here on the complexity of interactions between internal and external conditions in the shaping of perceptions over time. Among the diverse range of possible enquiry lines here I suggest a focus on at least three distinct clusters of societally defined human interests: *identity, order, niche*.[22] The assumption is that, throughout geography's disciplinary period, the nation-state has been the primary sponsor and audience for geographic research, and that during this period scholars have sought to justify their status by delivering products which enhanced national interests. This does not deny, of course, the potential universality of scientific horizons, the intellectual autonomy of individual scholars, or that of cross-national networks in academic thought and practice. To focus on these three sets of human interests seems particularly justifiable in the assessment of perception research, for two main reasons. First, they correspond to modes of environmental experience which philosophers claim are universal among humans: (1) the cognitive/symbolic (identity), (2) the interactional/behavioural (order), and (3) the organic/bio-ecological (niche).[23] Second, they afford excellent foci for research of the 'insider–outsider' phase; nations and managerial authorities may have quite different perceptions of identity, order, and niche, than those held by 'folk' from various regions, subgroups, and localities.

Mirrors, masks, milieux: each theme invites its own distinct set of research curiosities, but the most fascinating puzzle is how they interweave and combine. Together they help to elucidate, for example, how models developed in one area sometimes get applied to others – for example, geomorphological models applied to the study of urban dynamics, ecological models applied to social life, or systems dynamics applied to perceptual processes. Together they help elucidate the drama of migrants who carry their images with them to new environments, often encountering misunderstanding, success, or failure in establishing a new identity, and/or ecological nemesis resulting from the commitment to continue the practices of traditional *genres de vie*. Might this trilogy of themes not also serve as a comprehensive framework for evaluating the perceptual movement in geography?

First, it is perhaps beyond question that the most enduring influences on the development of a perceptual approach have come from pioneers like Humboldt, Granö, and Sorre, whose curiosities were about some 'whole picture' rather than about its constituent parts. Other early authors were concerned about how diverse elements combined to provide personality to a region or locality.[24] Some noted differences in folk 'mentality' between neighbouring regions,[25] others elaborated on the 'geographical lore' of particular periods.[26] Later on, researchers primarily concerned about the aesthetics of landscape appreciation have found that the perceptual element could afford key insight into cultural differences.[27] In all these cases, the cognitive element afforded key insight into the macro picture. Second, it is also fair to acknowledge that, in terms of analytical innovation and trend-setting, it was in the context of applied work or problem-solving that the relevance of perception studies was best recognized.[28] In hermeneutic vein, of course, one has to recognize that definitions of 'whole' as well as of 'problem' are inevitably contextual.

As with the traditional notion of *genre de vie*, this trilogy of themes affords not only a recipe for self-understanding, but also a set of lenses through which periods of research activity can be understood. What characterizes the contexts during which research interest in environmental perception has become evident? Through which doors have innovations found hospitality? A lot depends, no doubt, on the relative status of professional roles and institutions: much has to do with the nature and availability of intellectual expertise. Also, perhaps a lot more has to do with the political and social climate, national self-confidence, or self-questioning. A brief glimpse at the North American story may serve as illustration. Without even attempting an exhaustive survey, one could discern some broad patterns of reciprocity between scholarly practice and the general context of American life during periods in which the interest in environmental perception became evident.

Moments of cognitive renaissance: glimpses from America

Martyn Bowden suggests that there were three major waves of interest in environmental perception in Anglo-American geography: one in the 1920s, one in the 1940s, and a third, beginning in the 1960s and still continuing.[29] A brief look at two of these – the 1920s and the 1960s – may afford the opportunity to examine the interplay of mirrors, masks, and milieux in the formation of geographers' perceptions of reality.

The 1920s

Consider the 1920s in North America, a period during which geographers were busily establishing their disciplinary identity in a fairly indifferent milieu. High-ranking values must have been those of identity as an American profession, which implied a rejection of environmental determinism, a downplaying of European authorities, some internal democracy of effort, and above all some tangible evidence of potential relevance to practical problem-solving in the optimistic business-entrepreneurial spirit of Bowman's *New World*.[30] Aside from some brief flirtations with international boundary-definitions and debates over environmental determinism, human geographers sought a disciplinary identity and niche as American academics. They sought some status other than that of schoolteachers; there was need for some common language between physical and human branches of the field (for indeed, the geomorphologists still held greatest prestige). A science it was to be, of course, and for that one needed theories and models to guide analytical strategies. From which 'root metaphor' could such a guiding paradigm be derived?

Geography's world-view internationally, at least up to and during the First World War, had generally been an organicist one, bolstered by the progressive adoption of Darwinian ideas on evolution.[31] Prose of organicist flavour had succeeded in selling geography not only to educational authorities, but also to sponsors of frontier exploration on both sides of the Atlantic. To regard the earth as organism seemed to appeal also to the building of national spirit, regional identity, and historicist readings of culture: *pays*, *heimaten*, and 'natural regions' could be elegantly described as cells or nexuses within some larger whole. For ideological and other reasons, this world-view became unwelcome in North America after the First World War. Besides, where in all of this pioneering land could one find those *gemeinschaften* or *landschaften* for which the organism might be suitably applied? Certainly not on the Promethean frontier of 'civilizing rails' or on those vast expanses of resource territory towards which the imaginations of sponsors and audience were marching to conquer.[32] Besides, curiosities about 'natural areas' and 'ghettoes' were already amply expressed by Chicago sociologists, and rural sociologists could deal with agrarian communities. Geographers should cast their nets on wider expanses. For their part, they needed a world-view and analytical style which could allow for empirical field-work, survey, and inventory with a view to practical problem-solving. The prospect of being imprisoned in any overriding world-view such as 'the earth as organism', or any *a priori* formula for delivering a 'whole picture', however, was unsavoury. Enter Hettner's definition of the field as a chorological science: a formulation readily re-interpreted by Sauer and Hartshorne.[33] Chorology and morphology were to become the identifying mark of American geography throughout the late twenties and thirties: 'explore and map' transposed to 'inventory and prospect' surely fit the context of the Depression Years.

A rather lone voice in the crowd was John Kirtland Wright, a Harvard intellectual with training in history and a strong affection for European ideas. Though he must have felt quite at home with chorology and mapping, his world-view could scarcely be described as that of a formist. His philosophy appears to be much more attuned to that of William James and American pragmatism. His 'root metaphor' seemed closer to that of Pepper's contextualism: events, periods, places, each to be explored in terms of its own frame of reference. Indeed, one could argue that most of the twentieth-century initiatives for perception research in Anglo-America have had a strongly contextualist tone.[34]

Yet the *map*, metaphorically speaking, was to claim a more persistent hold on the consciousness of American geographers. Right through the 1960s, arguments for perception study were presented in terms of *mental maps*, *social space*, and even *behavioural space*.[35] Could it be that a basically chorological (or choropleth) approach enabled one to submit to the requisites of positivist method in documenting patterns? Even humanists, eventually, would stage their objections to positivism in terms of contrasts between *lived space* and *representational space*.[36] The root metaphor of *map* seemed to exercise an enduring hold, even among people who moved from the 'observation' to the 'insider–outsider' stances of enquiry.

Returning to the twenties and thirties, one could ask about which milieu interests might have inspired or steered the perceptions of geographers? Apart from a few Harvard intellectuals, the issue of national or regional *identity* was scarcely deemed a research-worthy issue. Of course the United States was the best-endowed of nations as far as natural resources were concerned.[37] Any possible identity problems of a demographic nature might easily be left to the sociologists and immigration authorities.[38] Predominant public interests, for which geography might have something to offer, were *order* and *niche*: the map provided an exceptionally effective grid-and-matrix within which information about resources, land use, urbanization, and transport could be documented. Inventory itself responded to an enduring public interest. Young recruits enjoyed summer-camp exercises in the apparently inexhaustible laboratory of the Midwest, convinced all the while that this work could be useful for the nation. Geography became a marketable skill, for treatises on the Corn Belt or covered wagons on the Santa Fe trail, for agricultural origins or

Maya landscapes, for Tennessee River Valley scenario or Dust Bowl salvage operations. And who better than they to raise questions about the wisdom of Federal environmental policy, for example on the Big Dams?[39] Who else might run the intelligence offices in Washington when war was in the air? Geographical facts, at least, could be readily inventoried and mapped, on the world projections and grid systems which had served so well at home. This was no time for speculating about images or perceptions; there was a war on: issues of domestic identity, order, and niche could scarcely compete with saving the world for democracy.

The 1960s

When the Second World War was over, North American geographers witnessed, among other changes, the rediscovery of another powerful metaphor of reality – mechanism. Throughout the 1920s and 1930s, their mentors had tended to discredit any talk about 'systems' as relics of European rationalism, quite inappropriate for the New World.[40] By the late 1940s, and especially in the 1950s (knights in shining armour now serving in Korea), the term 'system' regained esteem.[41] Wartime experience with operations research and images of the air age were no doubt already transforming the way geographers were beginning to perceive space and distance: orthographic projections were to outrival Mercator in basic atlases of the world.[42] Mechanism, a metaphor which opened the way for sophisticated modelling of target-bombing strategies, time–cost distance calculi, and technological innovation, eventually won geographers over from formal to functional approaches to their subject. Not only war veterans were involved in this transition from pattern to process.[43] Spatial interaction gained an edge on areal differentiation as key slogan for the discipline among many a young post-war recruit.[44] As the quantitative reformation reached its peak in the late 1960s, it seemed indeed that *map* and *mechanism* had found a mutually enriching *modus vivendi*: their combined insights could produce an impressive 'New Geography'.[45] Since both metaphors were analytically inclined, little effort was needed to exercise an appeal to the positivist spirit. Later on, when some researchers were to 'defect' from the certainties of a positivist (observational) stance on their objects of study, many still clung to the root metaphors of map and mechanism.

Early spokesmen for mental maps and environmental cognition could thus innovate and diversify methodological games within the categorical framework of formism and mechanism.[46] Murdie's

'Toronto' could examine the isomorphism (or lack thereof) between behavioural and perceptual spaces.[47] Comparisons and conflicts between objective and subjective levels of urban space could be captured within the framework of this most flexible metaphor.[48] For those who preferred a more rigorously 'systems' view of things – more commonly known as 'behaviourists' – connections between pattern and form were less interesting than were the dynamics of perceptual processes and their potential connections with overt behaviour.[49] For just as one has ascribed system-dynamics to the externally measurable processes of migration, shopping, producing, or consuming, so also one could infer mechanisms underlying perceptual processes, with ample justifications cited from precedents in psychology.[50] The rhetorical style and analytical routines of mechanism became a vogue which afforded a perceived common bond among researchers of diverse ideological orientation – positivist and anti-positivist, Marxist and anti-Marxist, bourgeois and anti-bourgeois, idealist and materialist – in the discothèque of behavioural geography.[51]

The most radical critique and alternative to the predominantly positivist approaches within environmental perception came from very different sources, for example from phenomenology, existentialism, and Marxism.[52] Issues raised were not simply those of knowledge foundations, or of the ethics involved in applied research; what was flashing on the horizon was the potential harvest of an 'insider–outsider' phase of enquiry. So what if one could spell out contrasts and conflicts between managerial and folk perspectives on life and landscape? What might the policy implications be, and to what extent could the geographer continue to be involved? The turmoil of the seventies separated idealist and materialist, Ivory Tower and market-place. It also invited *poesis*, and the acknowledgement that researchers were participants, not just observers of reality.

How was this to reverberate in perception research? Two important European sources of inspiration won special attention. Merleau-Ponty offered a definition of perception which included emotional, bodily, and linguistic aspects of human experience.[53] Alfred Schütz's 'mundane phenomenology' offered new possibilities for exploring everyday life.[54] Even more profoundly, the ideas of Heidegger and Gadamer brought about an awareness of language and interpretation: it was from these writers that the real impetus for Phase Three eventually came.[55]

Several curiosities should hopefully emerge from this rendering of the tale. Why, for example, were *map* and *mechanism* so favoured as root metaphors in the Anglo-American geographer's perceptions of reality? Why really did *organism* get overthrown? And why did contextualism – North America's major contribution to Western philosophy (and its geographic expression, *arena*) – become so poorly reflected in American geography until late in the 1970s? The first two, map and mechanism, have performed brilliantly in terms of *logos* and *ergon*; the latter seem to function best when the aims are those of *poesis* and *paideia*. Might this afford some rhetorical questioning about the perception movement and its possible sensitivity to societal context? Does it take war or peace, depression or prosperity, to call forth a 'cognitive renaissance' among practitioners of geography? And what does it take to encourage researchers to move from observer (spectator) to participant (experiencer) stance on the practice of geography?

Space does not allow further elaboration of these curiosities. Each tradition or school could sketch its own story and note the external circumstances – of war or peace, economic growth or decline, migration of people, and world events – which may have prodded the imaginations and channelled the energies of scholars in different periods. The North American experience does offer a reasonably good base from which to pose questions on the interweaving of mirrors, masks, and diverse milieux interests during phases of enthusiasm for research on environmental perception within the practice of geography.

Dream and reality of the 'behavioural environment'

Whither, then, geographers' explorations into the *terrae incognitae*? What future research horizons lie in store in the 'behavioural environment'? Assuming that its initial dream was to render better accounts of geographical experience, and/or to promote better ways of understanding the human use of the earth, has the reality of environmental perception research brought us closer to that goal?

Several lessons have been learned, and several new questions raised: on that basis alone geographers should applaud. Now we are at least aware that all humans (including ourselves) see, interpret, and articulate our environmental experiences in culturally specific ways. Our research products, too, are usually presented in discipline-specific languages. From this awareness, some dream of a language which could facilitate dialogue and mutual understanding between geographers and 'people'; others of a language which could elucidate the drama of interactions between humans and other life forms (animals, plants, birds, places) in the biosphere. This opens up a vast arena for *poesis*, far beyond the scope of this chapter, and only glimpses appear on the horizons, as in Lopez's treatise on the Arctic, or Matthiessen's pilgrimage to the Himalayas.[56] Valuable precedents among classical naturalists would be worth re-reading in this respect. On home ground, the exercise of seeking categories which could facilitate mutual understanding among researchers of different sub-specialties within our own discipline, or even among perception researchers in different countries, would merit some time and energy. One of the most demanding challenges posed by Phase Three, after all, is the critical reflection on one's own perceptions. At the risk of narcissism, one is challenged to an openness for the journey of self-understanding as precondition for mutual understanding, and eventually for understanding our human world. On the basis of selected facets of the North American record in research on environmental perception, then, certain questions might be raised about the 'dream and reality' of Kirk's post-war challenge.

The cognitive dimension in environmental perception

The Cartesian legacy, *cogito ergo sum*, and the Kantian *categorical imperative* have weighed heavily on the consciousness of geographers in the Western world. Heirs to a tradition in which intellect has been regarded as queen among human faculties, one has blithely assumed that human experiences of the world are fundamentally cognitive. Definitions of cognition have, of course, varied widely, depending on which psychologist one has consulted. And the psychological theories which have won greatest appeal among geographers have been those which accept a narrowly cerebral definition of cognition. Evidence mounts, however, that the understanding of ways in which other societies experience their environments may demand a much more nuanced definition of 'perception', and a sharper eye on how this may or may not be expressed in overt behaviour. This message comes most forcibly from researchers in environments quite different from those of the temperate zone, like Lopez's explorations in the Arctic:

> It is easy to underestimate the power of a long-term association with the land, not just with a specific spot but with the span of it in memory and imagination, how it fills, for example, one's dreams. For some people, what they are is not finished at the skin, but continues with the reach of the senses out into the land. If the land is summarily disfigured or reorganized, it causes them psychological pain. Again, such people are attached to the land as if by luminous fibers; and they live in a

kind of time that is not of the moment, but, in concert with memory, extensive, measured by a lifetime. To cut these fibers causes not only pain but a sense of dislocation.[57]

The lifeworld, even in Husserlian terms, comprises all of living creation – trees, lakes, animals, birds, as well as humans. Does not each realm of ongoing creation have its own mode of 'perceiving' environment? To wrench out the human part, or even to isolate its 'cognitive' part, inevitably disrupts the orchestra of lifeworld. The frontier therefore might well lie in allowing places, cultures, and biotic phenomena to express their own 'perceptions',[58] but also in acknowledging what overall mythological horizons might underlie our own 'perceptions'. In a brilliant account of landscape trans-formations in the Euro-American world during the past 200 years, for example, Berman has cited Goethe's Faust, *ein Menschens Geist in Seinen Hohen Streben* as key motif in human efforts to plan and develop cities and regions.[59] Umberto Eco discerns a return of the Middle Ages in today's museum-building, tamed nature reserves, and other paraphernalia of tourist landscapes around us.[60] To understand one's own perceptions of the world, then, one is obliged to uncover not only the responses to written questionnaires about housing, movement, and choice at any point in time, but also to delve into foundational myths which underlie our civilization's value choices. Berque's studies on Japanese landscape and civilization, Hanafi's studies on the Koran, Singh's narratives on the Ganges, and Bonnemaison's account of Vanuatu, all point in the direction towards which the perception movement might well be pointing.[61] For each of these worlds, geography could outline the challenge of self-understanding as portal to world understanding.

Symbolic transformations

Suzanne Langer, inspired in large part by Ernst Cassirer, claimed that the most characteristically human activity of all is not cognition, but symbol making.[62] Humans everywhere demonstrate creativity in transforming direct experience into symbols, for example via painting, sculpture, music, poetry, or mathematics. The West has witnessed a rich heritage of symbolic transforma-tions, from cave drawings and cathedral windows to the semiotics of engineering structures. Perhaps its very success explains why it has shown such little sensitivity to ways in which other societies have created theirs. Visual evidence ranks highest among sensory inputs to the Western cerebral process, but in several non-Western societies other senses – sound, smell, feeling, taste – may have held equally strong appeal as antennae on knowledge and environmental understanding.[63] Our analytical models of perception in the West have tended to fragment and ossify parts of sensory experience, submitting each to special examination. In that very process one may have truncated the prospect of establishing dialogue with researchers on perception in other civilizations. All of us are perhaps so imbued with the inherited myths of our respective cultures that we fail to recognize the challenge of exploring myth and metaphor – that realm of symbolic transformation which humanity has used to gain some kind of comprehensive understanding of its physical environment, and then to articulate insight on its own environmental experience.[64] In short, there are realms of environmental experience which may have escaped the analytical net of conventional models in Euro-American environmental perception research: the symbolic, the mythical, and the metaphorical. Recent work in semiotics may help redress this trend, or, on the other hand, it may push researchers back into an 'observation' stance on the world around.[65]

Once our colleagues in other lands succeed in transcending their well-justified anti-colonial and anti-imperialist feelings, and Euro-Americans become aware of the 'dream and reality' in their Faustian dreams, perhaps geographers all over the world might have a lot to share about coining a more experientially grounded definition of 'perception'.

Holistic versus fragmented perceptions

Western science has paid a high price for the power and acuity of its scientific procedures. Specialization has led to a fragmentation of thought and life, and reductionism has hitherto failed to inspire synthesis of any other kind than totalitarianism. Strictly speaking, a scientific observer sees 'nothing but' the realities screened into focus by particular research models. Submitting environmental perception to such *ceteris paribus* constraints, as some Euro-Americans have felt motivated to do, inevitably led to fragmented accounts of environmental experience. One aspect of this general puzzle could be noted here, namely the separation of space and time in analyses of the behavioural environment. American geographers opted for a Hettnerian recipe for disciplinary identity, with Newtonian notions of space as container of objects and events, at a time when the scientific world as a whole was clothing itself in an Einsteinian vision of space–time manifold as contextual definition of reality. Enthusiasts of 'process' tended to adapt a mechanical notion of space–time, examining 'spatial systems' of human interaction, or even perceptual processes at a moment in time. There was a grand promise of recapturing the temporal dimension in Hägerstrand's 'time-geography' model; it certainly heralded dynamism. Early work in time-geography, however, dealt primarily with clock or calendar (isometric) measures of time; it showed little interest in exploring the temporal sense of human perceptions. Yet via this brave endeavour, several new doors on environmental experience were implicit. Minkowski's *Lived Time* became the wellspring for curiosities about lived space;[66] problems of health and stress became interpretable in fresh ways by looking at the spatio-temporal context of everyday behaviour and memory.[67] In fact, it is now recognized by many a practising geographer that cross-cultural understanding cannot be achieved without a close examination of how different societies have seen, used, and valued time as well as space in their lifeworlds.[68]

The problem-solving imperative

Implicit in the foregoing account is the impression that perception research in America found appeal at moments when it could serve as (1) phoenix symbol of a renaissance, cognitive or otherwise, in professional practice, or (2) potential expertise for a Faustian attack on social or environmental problems. With a phoenix symbol, few would argue; with a Faustian banner many would commit themselves to 'make the system work at whatever price'; others would find ways of retreating to an 'observer' stance again via structuralist or semiotic interpretations. Focus on the problem-solving challenge, however, does allow for some critical evaluation of research on perception.

'Problems', as Schumacher observed, can be 'convergent' or 'divergent'.[69] There are those such as arranging the Lego parts of a shopping mall, designing a railroad network, which could be solved via calculi on throughput investments, costs, and benefits, and other cerebral exercises. These are the problems which Schumacher deems to be 'convergent'. But surely not all problems are construable in these terms. There are, for example, the gnawing worries about demographic changes, industrial obsolescence, regional decline, unemployment, or premature retirement, which are not easily construable in similar terms.[70] In the former ('convergent') case, the agenda may appear quite manageable; in the latter ('divergent') situation, quite a different approach might be in order. In much applied geography, with or without a 'perceptual' component, such distinctions have not always been recognized.

The problem-solving orientation (*ergon*) of perception research in North America has led to serious consequences as far as mirrors and masks are concerned. Over time, and especially with its scientific ambitions, perception research has migrated out of geography, and become the special domain of architects, planners, and psychologists. As far as mirrors are concerned, the contextualist approach was better equipped to describe events and situations than it was for defining policy for the management and control of environmental problems. Ironic indeed that, from thoroughly relativistic and inductive research, scholars have come forward with formulae for problem-management which have a decidedly imperialistic tone. The dilemma for anyone who wished to turn perception research into an instrument of planning policy was both philosophical and practical: ultimately it sought to maintain a relativistic stance when decisions often involved a clash of ideologies.

Perhaps it is essentially in this penchant for demonstrating the culturally relative and contextually contingent nature of all mirrors and masks that the basic lessons of perception research have emerged. Its enduring value, for whatever discipline that has adopted it, is ultimately the evoking of awareness about cultural differences in the experience of world. A vast horizon of exploration has been opened up. Each tradition faces its own challenge; each '-ism' and '-ology' would recount its own version of the story. The essential challenge which remains is that of communication. The ultimate harvest may depend on how well we manage to understand and be understood, not only by some folk-out-there, but also by colleagues on our own home ground.

Notes

1 D. Lowenthal (ed.) 'Environmental perception and behavior',
  Research Paper no. 109 (1967), Department of Geography, University
  of Chicago; J. Wolpert, 'Migration as an adjustment to environmental
  stress', *Journal of Social Issues*, vol. 22, no. 4 (1966), pp. 92–102;
  R. W. Kates and J. F. Wohlwill (eds), 'Man's response to the physical
  environment', *Journal of Social Issues* (special issue), vol. 22 (1966).
2 T. F. Saarinen and D. Seamon (eds), *Environmental Perception and
  Behavior: Inventory and Prospect* (University of Chicago Press,
  Chicago, 1984).
3 See M. J. Bowden, 'The cognitive renaissance in American
  geography: the intellectual history of a movement', *Organon*, vol. 14
  (1980), pp. 199–204; T. F. Saarinen and J. L. Sell, 'Environmental

perception', *Progress in Human Geography*, vol. 4, no. 4 (1980), pp. 525–48; T. F. Saarinen and J. L. Sell, 'Environmental perception', *Progress in Human Geography*, vol. 5, no. 4 (1981), pp. 525–47.

4   R. Geipel, 'La géographie de la perception en Allemagne Fédérale', *L'Espace Géographique*, vol. 3 (1978), pp. 195–8; P. Claval, 'L'évolution récente des recherches sur la perception', *Rivista Geografica Italiana*, vol. 87 (1980), pp. 6–24; A. Frémont (ed.), *Espaces vécus et civilisations*, Mémoires et Documents de Géographie (Centre National de la Recherche Scientifique, Paris, 1982); J. Gallais, 'La perception et la pratique de l'espace dans les pays tropicaux'; et 'La conclusion', in Frémont (ed.), *Espaces vécus et civilisations*, pp. 29–48.

5   J. Brunhes, *La Géographie humaine. Essai de classification positive* (Alcan, Paris, 1910, 1934); J. Sion, *Les Paysans de la Normandie orientale: Pays de Caux, Bray, Vexin Normand, Vallée de la Seine, étude géographique* (Armand Colin, Paris, 1909); G. Hardy, *La Geographie psychologique* (Armand Colin, Paris, 1939); E. Dardel, *L'Homme et la terre* (Presses Universitaires de France, Paris, 1952); M. Sorre, *La Géographie psychologique: l'adaptation au milieu climatique et biosocial* (Presses Universitaires de France, Paris, 1954); M. Sorre, *L'Homme sur la terre. Traité de géographie humaine* (Hachette, Paris, 1961).

6   A. Buttimer and T. Hägerstrand, 'Invitation to dialogue', DIA Paper no. 1 (1980), Kulturgeografiska institutionen, University of Lund; A. Buttimer, *The Practice of Geography* (Longman, London and New York, 1983); A. Buttimer and others (eds), 'Creativity and context', Report on a Symposium held at Sigtuna (June 1978), Lund Studies in Geography (1983), University of Lund; A. Buttimer, 'Ideal und Wirklichheit in der Angewandten Geographie', *Münchener Geographische Hefte*, no. 51 (1984).

7   P. Feyerabend, *Knowledge without Foundations* (Oberlin University Press, Oberlin, Ohio, 1961); R. Rorty, *Philosophy and the Mirror of Nature* (Princeton University Press, Princeton, NJ, 1979), pp. 315–94.

8   N. Bohr, 'On the notions of causality and complementarity', *Dialectica*, vol. 2 (1948), pp. 312–19.

9   T. Kuhn, *The Structure of Scientific Revolutions* (University of Chicago Press, Chicago, 1962); D. Stoddart (ed.), *Geography, Ideology and Social Concern* (Basil Blackwell, Oxford, 1981).

10 K. Boulding, *The Image – Knowledge in Life and Society* (University of Michigan Press, Ann Arbor, 1956); K. Lynch, *The Image of the City* (Harvard University Press, Cambridge, Mass., 1960).

11 R. Sack, 'Conceptions of geographic space', *Progress in Human Geography*, vol. 4, no. 3 (1980), pp. 315–45; K. R. Cox and R. G. Golledge (eds), *Behavioral Problems in Geography Revisited* (Methuen, New York and London, 1981).

12 H. Lefèbvre, *La Production de l'espace* (Editions Anthropos, Paris, 1974); A. Sayer, 'Can values be science free?', in D. T. Herbert and R. J. Johnston (eds), *Geography and the Urban Environment:*

Progress in Research and Applications, vol. 4 (John Wiley, Chichester, 1981), pp. 29–58; K. R. Cox, 'Bourgeois thought and the behavioral geography debate', in Cox and Golledge (eds), *Behavioral Problems in Geography Revisited*, pp. 256–79; D. Ley, 'Behavioral geography and the philosophies of meaning', in Cox and Golledge (eds), *Behavioral Problems in Geography Revisited*, pp. 209–30.
13 D. N. Livingstone and R. T. Harrison, 'Meaning through metaphor: analogy as epistemology', *Annals of the Association of American Geographers*, vol. 71 (1981), pp. 95–107; C. Rose, 'Wilhelm Dilthey's philosophy of historical understanding: a neglected heritage of contemporary humanistic geography', in Stoddart (ed.), *Geography, Ideology and Social Concern*, pp. 99–133; K. Christensen, 'Geography as a human science: a philosophical critique of the positivist–humanist split', in P. Gould and G. Olsson (eds), *A Search for Common Ground* (Pion, London, 1982), pp. 37–57.
14 E. Wirth, 'Zur wissenschaftstheoretischen problematik der länderkunde', *Geographische Zeitschrift*, vol. 66 (1978), pp. 241–61; G. Bahrenberg, 'Anmerkungen zu E. Wirth', *Geographische Zeitschrift*, vol. 67, no. 2 (1979), pp. 147–62; R. Mugerauer, 'Concerning regional geography as a hermeneutical discipline', *Geographische Zeitschrift*, vol. 69, no. 7 (1981), pp. 57–67.
15 S. C. Pepper, *World Hypotheses* (University of California Press, Berkeley, 1942).
16 A. Buttimer, 'Musing on Helicon: root metaphors in geography?', *Geografiska Annaler*, vol. 64, no. 2 (1982), pp. 89–96. Responses from various sources, particularly from cartographers, suggest that the terms 'form' or 'mosaic' might be more appropriate metaphors than 'map' to capture the world-view of formism.
17 G. H. Mead, *Mind, Self, and Society* (University of Chicago Press, Chicago, 1934); E. Goffman, *Interaction Ritual: Essays on Face-to-Face Behavior* (Anchor Books, Garden City, NY, 1967); A. Strauss, *The Social Psychology of George Herbert Mead* (University of Chicago Press, Chicago, 1956); A. Strauss, *Images of the American City* (Free Press, New York, 1961); J. S. Duncan and N. G. Duncan, 'Social worlds, status passage, and environmental perspectives', in G. T. Moore and R. G. Golledge (eds), *Environmental Knowing* (Dowden, Hutchinson & Ross, Stroudsburg, Pa., 1976), pp. 206–13.
18 E. L. Hughes, *The Sociological Eye. Selected Papers on Work, Self, and the Study of Society* (Aldine-Atherton, Chicago, 1971).
19 M. Teich and R. Young (eds), *Changing Perspectives on the History of Science* (Heinemann, London, 1973); S. Lilley, 'Cause and effect in the history of science', *Centaurus*, vol. 3 (1953), pp. 58–72; E. Mendelsohn and others, 'The social production of scientific knowledge', in *Sociology of Science Yearbook* (Reidel, Dordrecht, 1977).
20 J. Bernal, *Science in History* (C. A. Watts and Co., London, 1965); P. Bourdieu, *Outlines of a Theory of Practice* (Cambridge University Press, Cambridge, 1978).

21 Buttimer, 'Ideal und Wirklichheit in der Angewandten Geographie'.
22 Ibid.
23 G. Bachelard, *The Poetics of Space* (Beacon Press, Boston, 1958; trans. 1964); E. Cassirer, *Philosophy of Symbolic Forms* (Yale University Press, New Haven, Conn., 1955); Lefèbvre, *La Production de l'espace*.
24 P. Vidal de la Blache, *Tableau de la géographie de la France*, vol. 1 of E. Lavisse, *Histoire de France* (Hachette, Paris, 1903), translated as *The Personality of France* (Christophers, London, 1928); Sir C. Fox, *The Personality of Britain* (National Museum of Wales, Cardiff, 1932); E. E. Evans, *The Personality of Ireland. Habitat, Heritage and History* (Cambridge University Press, Cambridge, 1973).
25 Sion, *Les Paysans de la Normandie orientale*.
26 F. Braudel, *La Méditerranée et le monde méditerranéen à l'époque de Philippe II* (Armand Colin, Paris, 1949); J. K. Wright, 'Terrae incognitae: the place of the imagination in geography', *Annals of the Association of American Geographers*, vol. 37 (1947), pp. 1–15.
27 D. Lowenthal and H. Prince, 'English landscape tastes', *Geographical Review*, vol. 55 (1965), pp. 186–222; D. Linton, 'The perception of scenery', *Scottish Geographical Magazine*, vol. 84, no. 8 (1968), pp. 219–38; J. Appleton, *The Experience of Landscape* (John Wiley, London, 1975).
28 J. Brunhes, *L'Irrigation, ses conditions géographiques, ses modes et son organisation dans le péninsule Ibérique et dans l'Afrique du Nord* (C. Naud, Paris, 1902); G. F. White (ed.), *Natural Hazards: Local, National, Global* (Oxford University Press, London, 1974); Kates and Wohlwill (eds), 'Man's response to the physical environment'.
29 Bowden, 'The cognitive renaissance in American geography'; M. J. Bowden, 'Environmental perception in geography: a commentary', in Saarinen and Seamon (eds), *Environmental Perception and Behavior*, pp. 85–92.
30 I. Bowman, *The New World: Problems in Political Geography* (World Book Company, Yonkers-on-Hudson and Chicago, 1921); P. E. James and G. Martin, *The Association of American Geographers, Seventy Five Years, 1904–1979* (Association of American Geographers, Washington, DC, 1979).
31 Vidal de la Blache, *Tableau de la géographie de la France*; W. M. Davis, 'Geography in the United States', *Science*, vol. 17 (1904), pp. 121–32; A. J. Herbertson, 'The major natural regions: an essay in systematic geography', *Geographical Journal*, vol. 25 (1905), pp. 300–12.
32 M. Jefferson, 'The civilizing rails', *Economic Geography*, vol. 4 (1928), pp. 217–31.
33 C. O. Sauer, 'The morphology of landscape', University of California Publications in Geography no. 2 (1925), University of California, Berkeley, pp. 19–54; R. Hartshorne, *The Nature of Geography – A Critical Survey of Current Thought in the Light of the Past* (Association of American Geographers, Lancaster, Pa., 1939); P. E. James and C.

F. Jones (eds), *American Geography: Inventory and Prospect*
(Syracuse University Press, Syracuse, NY, 1954); James and Martin,
The Association of American Geographers, Seventy Five Years.
34 Wright, 'Terrae incognitae'; W. Kirk, 'Historical geography and the
concept of the behavioural environment', *Indian Geographical
Journal*, Silver Jubilee Volume (1952), pp. 152–60; D. Lowenthal,
'Geography, experience, and imagination: towards a geographical
epistemology', *Annals of the Association of American Geographers*,
vol. 51 (1961), pp. 241–60; H. C. Brookfield, 'On the environment as
perceived', *Progress in Geography*, vol. 1 (1969), pp. 51–80.
35 P. Gould and R. White, *Mental Maps* (Penguin, Harmondsworth,
1974); A. Buttimer, 'Social space in interdisciplinary perspective',
Geographical Review, vol. 59 (1969), pp. 417–26; R. A. Murdie,
'Factorial ecology of metropolitan Toronto, 1951–1961. An essay on
the social geography of the city', Department of Geography Research
Paper no. 116 (1969), University of Chicago.
36 Yi-Fu Tuan, 'Geography, phenomenology, and the study of human
nature', *Canadian Geographer*, vol. 15 (1971), pp. 181–92; M.
Samuels, 'Science and geography: an existential appraisal',
unpublished Ph.D. thesis, Department of Geography, University of
Washington, Seattle, 1971.
37 Davis, 'Geography in the United States'; J. Russell Smith, *North
America, its People and the Resources, Development, and Prospects of
the Continent as an Agricultural, Industrial, and Commercial Area*
(Harcourt, Brace & Co., New York, 1925); Jefferson, 'The civilizing
rails'.
38 See, however, E. Huntington, *The Human Habitat*, 4th edn (Van
Nostrand Co., New York, 1927).
39 W. Kollmorgan, 'And deliver us from the Big Dams!', *Land
Economics*, vol. 30 (1954), pp. 333–46.
40 J. Leighly, 'Some comments on contemporary geographic methods',
Annals of the Association of American Geographers, vol. 27 (1937),
pp. 127–41.
41 L. von Bertalanffy, 'General systems theory: a new approach to the
unity of science', *Human Biology*, vol. 23 (1951), pp. 303–61;
Boulding, *The Image – Knowledge in Life and Society.*
42 P. M. Morse and G. E. Kimball, *Methods of Operations Research*
(John Wiley, New York, 1951).
43 James and Jones (eds), *American Geography: Inventory and Prospect;*
J. Blaut, 'Space and process', *Professional Geographer*, vol. 13 (1961),
pp. 1–7.
44 E. A. Ullman, 'Geography as spatial interaction', *Annals of the
Association of American Geographers*, vol. 44 (1954), pp. 283–4;
Buttimer, *The Practice of Geography*, pp. 186–95.
45 R. J. Chorley and P. Haggett, *Frontiers in Geographical Teaching*
(Methuen, London, 1965); R. J. Chorley and P. Haggett, *Models in
Geography* (Methuen, London, 1967); D. Harvey, *Explanation in
Geography* (Edward Arnold, London; St Martin's, New York, 1969);

W. K. D. Davies, *The Conceptual Revolution in Geography* (Rowan & Littlefield, Totowa, NJ, 1972).

46 Lynch, *The Image of the City*; R. M. Downs and D. Stea, *Image and Environment: Cognitive Mapping and Spatial Behavior* (Aldine, Chicago, 1973); Gould and White, *Mental Maps*.

47 Murdie, 'Factorial ecology of metropolitan Toronto'.

48 P. H. Chombart de Lauwe and others, *Paris et l'agglomération parisienne* (Presses Universitaires de France, Paris, 1952); R. Ledrut, *L'Espace social de la ville. Problèmes de sociologie appliqués à l'aménagement urbain* (Editions Anthropos, Paris, 1968); M. J. Bertrand and A. Metton, 'Méthode d'étude géographique du milieu résidentiel', *L'Information géographique*, vols 2 and 3 (1975); A. Metton, 'Le quartier: étude géographique et psychosociologique', *Canadian Geographer*, vol. 13, no. 4 (Winter 1969), pp. 299–316.

49 G. Rushton, 'The scaling of locational preferences', in K. R. Cox and R. G. Golledge (eds), 'Behavioral problems in geography: a symposium', Studies in Geography no. 17 (1969), Evanston, Ill., pp. 197–227; Moore and Golledge (eds), *Environmental Knowing*.

50 K. Lewin, *A Dynamic Theory of Personality* (McGraw-Hill, New York, 1935); W. Köhler, *Gestalt Psychology* (Bell & Sons, London, 1930); Kates and Wohlwill (eds), 'Man's response to the physical environment'.

51 Cox, 'Bourgeois thought and the behavioral geography debate'; Ley, 'Behavioral geography and the philosophies of meaning'.

52 D. Seamon, *A Geography of the Lifeworld* (Croom Helm, London; St Martin's, New York, 1979); D. Seamon, 'The phenomenological contribution to environmental psychology', *Journal of Environmental Psychology*, vol. 2 (1982), pp. 119–40; Yi-Fu Tuan, 'Geography, phenomenology, and the study of human nature'; Yi-Fu Tuan, *Topophilia: A Study of Environmental Perceptions, Attitudes and Values* (Prentice-Hall, Englewood Cliffs, NJ, 1974); E. Relph, *Place and Placelessness* (Pion, London, 1976); J.-P. Ferrier, B. Racine, and C. Raffestin, 'Vers un paradigme critique: matériaux pour un projet géographique', 1, *L'Espace Géographique*, vol. 4 (1978), pp. 291–7; N. Smith, 'Geography, science and post-positivist modes of explanation', *Progress in Human Geography*, vol. 3, no. 3 (1979), pp. 356–83; A. Buttimer, 'Grasping the dynamism of lifeworld', *Annals of the Association of American Geographers*, vol. 66 (1976), pp. 277–92.

53 M. Merleau-Ponty, *The Phenomenology of Perception* (Northwestern University Press, Evanston, Ill., 1962).

54 A. Schütz, *Collected Papers*, 2 vols. (Martinus Nijhoff, The Hague, 1962); A. Schutz, *On Phenomenology and Social Relations: Selected Writings*, ed. H. Wagner (University of Chicago Press, Chicago, 1970).

55 M. Heidegger, *Vorträge und Aufsätze* (Neske, Pfullingen, 1954); M. Heidegger, *Poetry, Language, Thought* (Harper & Row, New York, 1971); H. F. Gadamer, *Wahrheit und Methode*, 2nd edn (J. C. B.

Mohr, Tubingen, 1965), translated as *Truth and Method* (Seabury Press, New York, 1975).
56 Buttimer and Hägerstrand, 'Invitation to dialogue'; B. Lopez, *Arctic Dreams: Imagination and Desire in a Northern Landscape* (Charles Scribner's Sons, New York, 1986); P. Matthiessen, *The Snow Leopard* (Bantam Books, New York, 1970).
57 Lopez, *Arctic Dreams*, p. 279.
58 See ibid.; T. Schwenk, *Sensitive Chaos* (Shocken, New York, 1976); A. T. de Nicolas, *Meditations through the Ṛg Veda* (Shambala, London, 1978).
59 M. Berman, *All that is Solid Melts into Air* (University of California Press, Berkeley, 1982).
60 U. Eco, *Travels in Hyper-reality* (Harcourt, Brace, Jovanovich, New York, 1986).
61 A. Berque, *Vivre l'espace au Japon* (Presses Universitaires de France, Paris, 1982); Hassan Hanafi, 'Human subservience of nature: an Islamic model', SALFO Symposium on Natural Resources in Cultural Perspective (May 1982), SALFO, Stockholm, 1982; R. P. B. Singh, 'Clan settlements in the Saran Main (Middle Ganga Valley): a study in cultural geography', Research Publication no. 18 (1977), The National Geographical Society of Inida, Varanasi; J. Bonnemaison, 'The tree and the canoe', *Pacific Viewpoint* (special issue), vol. 25, no. 2 (October 1984), pp. 117–51.
62 S. Langer, *Philosophy in a New Key* (Harvard University Press, Cambridge, Mass., 1957); Cassirer, *Philosophy of Symbolic Forms*.
63 de Nicolas, *Meditations through the Ṛg Veda*.
64 Bonnemaison, 'The tree and the canoe'.
65 F. Choay, 'Semiologie et urbanisme', *Architecture d'Aujourd'hui*, vol. 153 (1967), pp. 8–10; K. Foote, 'Space, territory and landscape: the borderlands of geography and semiotics', *Recherches Semiotiques/ Semiotic Enquiry*, vol. 5 (1985), pp. 158–75.
66 E. Minkowski, *Lived Time: Phenomenological and Psychopathological Studies* (Northwestern University Press, Evanston, Ill., 1933); O. Bollnow, 'Lived space', *Philosophy Today*, vol. 5 (1961), pp. 31–9; Yi-Fu Tuan, *Place and Space: The Perspectives of Experience* (University of Minnesota Press, Minneapolis, 1977).
67 A. Reinberg *et al*, *L'Homme malade du temps* (Stock, Paris, 1979); M. Melbin, 'Night as frontier', *American Sociological Review*, vol. 43 (1978), pp. 3–22.
68 J. Galtung, 'Sivilisasjon, kosmologi, fred og utvikligg', *Det Norske Videnskapsakadem Arsbok* (1981), pp. 130–53; S. Suri, 'Role of small and intermediate size cities in national development', United Nations Centre for Regional Development, Nagoya, Japan, 26 January – 1 February 1982; S. Suri, 'The crowd in India', *Illustrated Weekly of India*, 23 January 1983, p. 55; J. T. Fraser and N. Lawrence (eds), *The Study of Time*, 2 vols (Springer-Verlag, New York, 1975, 1978).
69 E. F. Schumacher, *A Guide for the Perplexed* (Abacus, London, 1978).

Re-evaluation

70 C. S. Davies, 'Wales: industrial fallibility and spirit of place', *Journal of Cultural Geography*, vol. 5 (1985), pp. 72–86; C. S. Davies, 'the throwaway culture: job detachment and rejection', *Gerontologist*, vol. 25, no. 3 (June 1985), pp. 228–31.

Chapter fourteen

A curiously unbalanced condition of the powers of the mind: realism and the ecology of environmental experience

Edward Relph

There is a widely held assumption that sense perceptions are a sort of dirty window between us and reality, obscuring the way the world actually is. Beyond this obscuring screen of fallible perceptions and subjective impressions there is believed to be a reality of enduring substances and processes that can best be revealed by the methods of objective research.

I consider this assumption to be unwarranted. It is defensible neither by an appeal to direct experience nor by an appeal to reason. It has nevertheless persisted for several centuries and has penetrated most academic disciplines, possibly because it allows those who claim to be able to grasp the 'real environment' beyond the façade of perceptions to make themselves into gatekeepers and to arrogate to themselves the authority which comes with access to privileged knowledge. Among its many consequences is the belief that anything to do with immediate experience, including judgements about quality of environments, is personal and secondary.

I take what I see and otherwise sense seriously. It is not a smokescreen between me and the objectively determined world of atoms or economic systems or whatever. One of my great interests is in landscapes. The landscapes I perceive are a primary reality: there is nothing more real above, below, beyond, or behind them. This does not mean that I doubt the existence of larger and smaller patterns and processes, for instance those of the hydrological cycle, institutional relationships, or atomic structures, but these have no more substance than my perceived world: indeed, to the extent that they are imaginative abstractions devised to account for measurements they have considerably less substance than my perceptions.

The assumption that what is known of environments consists only of sense perceptions and the ideas we form about these perceptions has important implications. From this perspective the senses are understood to be creative and active; they bind us to environments in an ecology of wonder and experience. They can be educated, made more discerning and discriminating. If you believe that the perceived world is just a dirty window then there would be little point in doing this, or indeed in making judgements about the quality of the dirtiness. If you accept, however, that there is no better reality than that given to us by our perceptions, then through the education of the senses it becomes possible to create the foundations for doing something which seems almost to have vanished from our understanding of environments – to acknowledge that beauty is not merely in the eye of the beholder and that there are indeed shared grounds for recognizing and perhaps creating excellence in places.

Realism and the world of perception

It could be that the habit of distinguishing a world of perceptions and a world of substance has ancient origins. Be that as it may, the clearest modern formulation, and the one which informs so much current thought, is probably John Locke's argument, made in the late seventeenth century in the light of Descartes's philosophy and Newton's science. He proposed that there are two types of qualities which we perceive in things. 'First, such as are utterly inseparable from the body in what state soever it be – solidity, extension, figure, mobility'. These he called original or primary qualities. 'Secondly, such qualities which in truth are nothing but powers to produce various sensations in us . . . as colours, sounds, tastes, etc.' These he called secondary qualities.[1]

Primary qualities for John Locke were real and existed in objects whether 'anyone's senses perceive them or no'. For secondary qualities, however, 'take away the sensation of them, let the eyes not see light or colours, nor the ears hear sounds, let the palate not taste, nor the nose smell, and all colours, tastes, odours and sounds . . . vanish and cease'.[2] One might, perhaps, wonder just how these are known to vanish in the absence of any sensations, but this was not a concern which Locke chose to address. For him it was self-evident that there is a real world composed of primary qualities which exists independently of humans, and another insubstantial world of secondary qualities given by the senses. This was consistent with what was apparently then being disclosed by scientists as they dug through surface forms and impressions to reveal the structure of matter and the laws of nature.

The assumption that the real world lies behind (or below, or beyond) the environments of everyday experience has become deeply and widely entrenched in science and social science. The philosophical name for this assumption is realism. Lenin was a confirmed realist. In 1908 he wrote:

> The realism of any healthy person who has not been an inmate of a lunatic asylum, or a pupil of one of the idealist philosophers, consists in the view that things, the environment, the world, exist independently of our sensation, of our consciousness, of our self and of man in general.[3]

Presumably for Lenin the pre-eminent lunatic would have been the first idealist philosopher and profound critic of John Locke's argument, George Berkeley.

In the first decade of the eighteenth century, Berkeley reflected carefully on Locke's assumption that the real world exists apart from the senses, and found he could not accept it. He wrote:

> The reality of things consists in being perceived. Light and colours, heat and cold, extension and figure, in a word the things we see and feel, what are they but so many sensations, notions, ideas or impressions on the sense; and is it possible to separate, even in thought, any of these from perception? For my part I might as easily separate a thing from itself.[4]

The difficulty Berkeley exposed was that what we know of things we must know through our senses and ideas formed on the basis of sense impressions. Matter stripped of all sensible qualities can be neither apprehended by the mind nor perceived by the senses. If there is a material substratum to existence it can never be known.

This argument of Berkeley's quickly established itself as idealism – the school of thought which is, rather misleadingly, held to maintain that environments exist in our ideas, in our minds. It did not meet with great favour among Berkeley's contemporaries, though it proved remarkably difficult to refute. Samuel Johnson, reminded of this by Boswell, petulantly kicked a stone and exclaimed 'I refute it thus!' He was wrong. This was rather a confirmation of Berkeley's argument because the sight of the stone and its movement, and no doubt the pain in his toe, were all too clearly sensations. Berkeley was, in fact, explicit that the sun, the stars, rivers, houses, mountains, trees, and even stones retain their reality under his perspective.

> Whatever we see, feel, hear, or anywise conceive or understand remains as secure as ever, and is as real as ever. . . . The only thing denied is 'matter' – this supposedly solid world of mass and extension which is not a part of our experience but a philosophical or scientific substance – a fabrication of the mind.[5]

This is also a view which has its twentieth-century supporters, one of them at least from what might seem to be an unlikely field – physical science. Erwin Schrödinger was one of the major contributors to the development of quantum physics, and his discoveries in the field of wave mechanics led him to the conviction that we do not live in two worlds, one perceived and one real.

> If, without involving ourselves in obvious nonsense, we are going to be able to think in a natural way about what goes on in a living, feeling, thinking being ... then the condition of our doing so is that we think of everything that happens as taking place in our experience of the world, without ascribing to it any material substratum.[6]

The external world, he argued, is known through sense-impressions. These are different for each individual; correspondence between them is established through language and not by reference to some underlying reality of primary qualities. Indeed, the distinction between primary and secondary qualities Schrödinger considered to be absurd, comparable to a belief that

> somewhere among the intergalactic nebulae there is a select spot where man-like beings with wings, in long white robes, are engaged in making sweet music and enjoying paradise. In both cases the burden of proof rests on the defenders of so bizarre an assertion.[7]

For the realist idealism is a form of lunacy, for the idealist the assumptions of realism are a bizarre fantasy.

Rejecting realism

What does this dispute about the nature of reality between authorities from seventeenth- and eighteenth-century philosophy, and twentieth-century Marxism and science, have to do with humble citizens or mere geographers? A great deal, because the philosophical assumptions of realism have insinuated themselves deeply into thought and practice. They inform much discussion in economics, in behavioural geography, and in most forms of abstract social and physical science. It seems to be generally accepted that there are lesser and greater forms of reality – the lesser one consisting of surface appearances and everyday experiences, the greater one consisting of more or less obscure measurements, observations, and models of underlying processes and structures. Richard Chorley, for example, has declared that geomorphologists must penetrate the appearance of external forms to reach the essential chemical and physical processes which sustain them.[8] Similarly, in William Kirk's account of the geographical environment an explicit distinction is drawn between the perceptual or behavioural world and the phenomenal world which underpins it.[9] Actually, I understand Kirk's account of experience to be an important first step in geographical thought away from simplistic realism to a recognition that perceptions do colour environments. My criticism of it is that it does not go far enough and reject entirely the idea of the phenomenal environment. In subsequent work in behavioural geography the usual aim has been to clarify the perceptual world, in some cases for its own sake, but in other cases apparently to explain why people do not behave in the rational manner of the real world beyond. So long as the assumption of an underlying objective reality remains, this aim constitutes a commitment to understanding something that is secondary both in its qualities and its relative importance.

The consequences of the realist perspective are more than academic. Thus, for instance, rationalistic, abstract knowledge, based implicitly or explicitly on assumptions about an objectively existing world, has become the preferred foundation for much planning and design. The process seems to work like this: the environments of everyday experience are compared with the supposed underlying order of things, and are found wanting; it is then but a short step to remaking the world to correspond with the image of abstract reality. This process can be discerned in the adoption of economic policies based on monetarist theories, in public-housing developments of great slab apartments laid out in rows which correspond with the tenets of modernist housing theory, and in new-town plans like the roundabout-infested grids of Milton Keynes and Telford which were devised to accord with theoretical notions about neighbourhoods, mobile populations, and communities without propinquity.

This tendency to remake the world to accord with an ideal image is not, of course, a recent invention. Indeed, Arthur Koestler argued that throughout history there has been a persistent desire to conform to some ideal and to belong to some whole larger than the individual, to God, to king, to country, to communism, to capitalist democracy.[10] Such willing conformity to abstract ideals has been a major source of violence on the one hand and subjugation on the other. The distinctive modern version of it is based on development of and subservience to scientific expertise. Koestler discussed at some length the research of the psychologist Stanley Milgram on obedience to authority. Milgram devised an experiment in which ordinary men and women were asked to inflict apparently real electric shocks on human subjects (in fact, actors trained to fake appropriate levels of agony).[11] Most of the participants were willing to cause extreme suffering, especially when they were reassured that this was necessary for the sake of science, even though they could see the actors in pain. It seems that people will accept and even participate in destructive processes if they feel absolved of responsibility, and the disclosure of an abstract reality through the methods of science is one important source of such absolution.

It appears that what has happened is that those promoting and implementing rationalistic plans and projects have been so far drawn into the myth of objective reality that they have often lost contact with actual places, and with real individuals and their concerns. They then use their authority and expertise to persuade others less knowledgeable to join them in the great effort to make their illusions and bizarre assertions real. Geographers no less than other professionals are guilty of this, of using expertise to promote theoretically more efficient settlement patterns or to devise systems of warfare aimed at destroying societies by attacking the geographical interdependencies of people and land.[12]

The assumptions of realism and their attendant consequences I find unacceptable. The existence of a material substratum of reality can be posited only through a leap of faith. It is a leap I am unable to make. The only reality I know is the one I perceive and think about on the basis of my perceptions. It is composed of things and events seen and experienced in all their multicoloured, obtuse particularity. It is nevertheless possible to understand this immediately experienced world either as a small part of larger-scale processes and patterns, or as a large manifestation of countless smaller processes and structures. This extension is possible because memory enables the comparison of immediate experiences through time, language makes possible the correspondence of sense impressions between individuals, and various mechanical devices such as aerial photographs and electron microscopes enlarge the range of the senses. Abstract thought, generalization, and imagination, for instance in models of spatial diffusion and DNA codes, also apparently extend the capacities of the senses, though it must be remembered that the products of these faculties are always mental constructs to explain or elaborate perceived realities. In none of these cases of extension and comparison do I suppose for a moment that immediate perceptions have been displaced by a new, better, or independently existing reality. The directly perceived world remains, although my appreciation of it may have been enhanced by my ability now to grasp it as a part of a larger or smaller set of processes. John Ruskin, as always, had the right words for it: 'Every advance in the acuteness of our perception will show us something new; but the old and first discerned thing will still be there, not falsified, only modified and enriched by the new perception.'[13]

This view of things can be called radical, empirical, or phenomenological, or what you will; the name is of no great significance. Since it owes some of its form to Berkeley it could even be called idealist, but that is misleading because it has to do with directly perceived realities and it is manifestly the case that it is 'realism' which is lost in ideas and abstract illusions. It is for those who trust their own senses and who prefer neither to subjugate others to expertise nor to have their own lives determined by the assumptions and theories of self-appointed authorities. Its deficiencies are perhaps a tendency to aestheticism and an overemphasis on particulars that can lead to parochialism. In spite of these weaknesses it suggests a different and, I believe, gentler understanding of the relationships between people and environments that can begin to counterbalance the bizarre assertions of realism and the imposition of abstract models on particular places.

Notes on the ecology of environmental experience

Given that environments are as they are perceived, it follows that the senses are bonds between people and their worlds. Edith Cobb wrote about an ecology of imagination, a continuum of mind, culture, and nature, the give and take between inner and outer worlds.[14] The very use of the term 'ecology' in this way is provocative. It suggests that each of us is connected to the world through a complex ecology of wonder, activity, sensation, memory, and vicarious experience. Our senses connect us to environments. This process is active, creative, and adaptive. From a realist point of view senses are boring things, as dull and passive as satellite dishes. However, perception is more than a mechanical process: we move around, sensing different complexes of things at each moment, sometimes attending to sounds, sometimes to sights, interpreting these, learning from our experiences, comparing them with memories, imbuing objects and landscapes with meanings.

The ecology of environmental experience has many processes and components. Cobb writes about wonder and imagination as means of organizing perceptions, especially in childhood. Language too must be important in this. There are questions of mood and short-term shifts in experience, of conviction, the role of memory, prejudices which emphasize one sense at the expense of the others, the influences of social institutions, the role of habit, of education, of greed and striving for power, the character of illusions and deceptions, of self-deceptions. A fully articulated ecology of environmental experience would have to elaborate and relate all of these. Here I will deal briefly with just one matter that connects closely to the problems of realist abstraction – implication and detachment.

In everyday life we are not primarily spectators; we are engaged with things. This engagement has several distinguishable forms and intensities. There are situations in which I am completely involved when the words 'environment' and 'I' cease to have meaning. This might happen when I am part of a crowd at a rally or a soccer game, writing an examination, climbing a mountain, swimming, whenever I am absorbed in what I am doing. In such cases the self is literally forgotten. Thus Bernard Berenson wrote in his autobiography that 'I never enjoyed to the utmost ... a landscape without sinking my identity into [it], without becoming it'.[15] Such unselfconsciousness and profound implication in environments is rarely more than fleeting. I get uncomfortable, or anxious about the time, or consider photographing the landscape: my individuality is recalled and thinking interjects itself between me and my world.

Most environmental experience consists of these two states of involvement – the former characterized by what might be called enfoldment, in which thoughts submit to sensations, the latter by reflection, in which sensations are suppressed by thinking. Usually the two states alternate and this maintains a balance between them. However, when seeing or one of the other senses becomes dominant it can lead to a self-centred striving for impressions and sensations. Conversely, thinking can gain at the expense of perception in three distinct ways. Compassionate objectivity is the detachment of the naturalist identifying plants or birds out of a fascination with their intrinsic characteristics, or a mother watching the behaviour of her child; it is simultaneously disinterested yet filled with affection. Arrogant objectivity self-consciously sets us outside environments, 'to make ourselves masters and possessors of nature' as Descartes had it.[16] This latter attitude arrogates one aspect of the ecology of environmental experience above all the others: it is closely linked to the problems of realism. The third way in which thinking can assert itself is unwilled and unwanted, the feeling of apartness that comes upon us in moments of depression, rendering the world absurd and empty of meaning. 'I am in the midst of things, nameless things. Alone without words, defenceless, they surround me, are beneath me, are behind me, above me. They demand nothing, they don't impose themselves; they are there.' So Sartre had Roquentin write in his diary in *Nausea*.[17]

Arrogant, deliberate objectivity tends to be destructive of environments and of others; existential objectivity is destructive of self. In both cases the ecology of environmental experience is shattered as the processes of perception are sundered or polluted beyond recognition. In contrast, compassionate thought reinforces the bonds of perception.

The practical challenge from my radical empirical perspective is to maintain the connections and to keep the balance between sensing and thinking. This is what I suppose Ruskin to have meant when he wrote: 'A curiously balanced condition of the powers of the mind is necessary to induce full admiration of any natural scene'. And again, 'If the thought were more distinct we should not see so well; and beginning definitely to think, we must comparatively cease to see.'[18] This is not to disdain thinking and objectivity but to keep them in perspective. Since abstraction and arrogant objectivity seem now to be in the ascendancy, especially in academic circles, what this actually means is that it is necessary to cultivate the powers of perception. This is a task somewhat out of fashion; but as anyone knows who has ever painted or drawn or played a musical instrument, or merely bothered to look carefully at things for a while, it is entirely possible to make at least some of one's senses more observant and discriminating. It is possible but not easy. Improving the sensitivity of one's perceptions demands discipline and practice, often without immediate or tangible rewards. There is no handbook on how to do this though one can learn from artists and authors who have themselves followed this route. From Goethe, perhaps, who set out on his tour of Italy determined 'to see with clear, fresh eyes'; or Ruskin, for whom careful looking at buildings and landscapes was a duty; or the paintings, sketches, and notebooks of the abstract artist Paul Klee. This is by no means a flawless process and it is well always to remember Ruskin's warning that 'One of the most singular gifts, or, if abused, most singular weaknesses, of the human mind is its power of persuading itself to see whatever it chooses.'[19]

The problem of quality

One of my chief interests is in the character and quality of landscapes. That it is not much in fashion to describe modern urban landscapes in terms of excellence is, I suspect, partly because so little in these commends itself obviously to the senses, partly because judgements of quality require a confidence that is not easily achieved without educating the senses, and partly because in a culture committed to realism such judgements are of little consequence. The qualities of landscapes could perhaps be assessed by a survey of opinions, but the likelihood is that this would lead to a mediocre average. Authorities could be consulted, but their views are likely to be narrow and self-serving.

The qualities of landscapes are apparent for anyone who makes the effort to look carefully and compassionately, with what Iris Murdoch has called 'unpossessive contemplation',[20] and is then willing to trust the evidence of their own eyes. Under such scrutiny good landscapes are those which reveal evidence of the expenditure of some personal or local effort and commitment in their making and maintenance. Whatever has been done easily, as it were by rote or by mindlessly following models for neighbourhood planning and apartment construction, will be at best mediocre. Good landscapes have signs of generosity, elements which serve no great economic or technical purpose but which can give pleasure; these include flower gardens in front yards, outdoor cafés, steps where people can sit in the sun. Poor landscapes are regimented, the edges are hard, the lines are straight, buildings are in rows, signs forbid small pleasures such as walking on the grass, abstraction has been imposed. Good landscapes are imperfect in small ways because they are a product of human efforts, and reflect human failings as well as human abilities. Everything does not fit together perfectly, the colours and setbacks and fenestration patterns are not all co-ordinated. Poor landscapes are those which seek perfection and inevitably fail to attain it; one need only think of the public-housing disasters which were made to be modern and egalitarian and which have had to be demolished. Good landscapes show signs of responsibility to whatever has been inherited from the past and what will be bequeathed to the future. They neither preserve nor erase everything old; they will age and change. Poor landscapes either have an invented history or no history and no future because abstract ideals are outside lived time. Good landscapes are those in which the hard and dirty work, and the clean and pleasant work, are reasonably shared by all. Ugly landscapes are found wherever the abstract science of getting rich for some (Ruskin's definition of economics) has created for others the palpable reality of being poor.

The recognition of landscape qualities such as these does not require a formal education, nor great sums of money to be spent on research, nor even being a member of one of the more leisurely classes. It does demand that a widespread lethargy of perception be overcome, a willingness to look with clear fresh eyes, the confidence to trust one's own observations, and perhaps the compassionate knowledge that comes with participating in the making and maintenance of a particular place. These experiences should be available to everyone.

Out of such an effort of observation and assessment should emerge not a handbook of standards for judging landscapes and environments, but informed opinions about qualities. There will, of course, be differences of opinion. These can provide the foundation for a debate which has to do with more than technical rearrangements, more than ideological confrontations, more than economic efficiency, because it is based on an understanding of the rich ecology of environmental experiences. Realism and excessive rationalism deny excellence because they make all questions of quality into secondary concerns. The quality of landscapes and environmental experience is too important to be left to those who assume that reality lies not in perceptions and immediate involvement with particular places but in some forever inaccessible substratum of existence.

Acknowledgements

I wish to thank Engin Isin for the quotation from Lenin and for an enlightening conversation about Locke and Berkeley; and Wayne Reeves for his research into Ruskin's sense of landscape.

Notes

1 J. Locke, *An Essay concerning Human Understanding*, ed. P. H. Nidditch (Clarendon Press, Oxford, 1975; first published 1690), ch. VIII, sections 9 and 10.
2 Ibid., ch. VIII, section 17.
3 V. I. Lenin, *Materialism and Empirio-criticism* (Foreign Languages Press, Peking, 1972; first published 1908), pp. 68–9.
4 G. Berkeley, *A New Theory of Vision* (Dent, London, 1954; first published 1709), part 1, section v.
5 Ibid., section xxxiv.
6 E. Schrödinger, *My View of the World* (Cambridge University Press, Cambridge, 1964), pp. 66–7.
7 Ibid., pp. 90–1.
8 R. J. Chorley, 'Bases for theory in geomorphology', in C. Embleton, D. Brunsden, and D. K. C. Jones (eds), *Geomorphology: Present Problems and Future Prospects* (Oxford University Press, Oxford, 1978), pp. 9–10.
9 W. Kirk, 'Problems of geography', *Geography*, vol. 48 (1963), pp. 357–71; see especially figure 6. Since much of my argument is almost opposite to that of Kirk's and since this is a volume in his

honour, I must note that Kirk's ideas helped to form my interests in phenomenology in the late 1960s, at which time I read them as an important corrective to simplistic realism in geography. While I may criticize them now for not going far enough, I am indebted to them as a point of departure for connecting philosophical arguments with environmental knowledge and the concerns of geography.

10 A. Koestler, *Janus* (Random House, New York, 1978).

11 S. Milgram, *Obedience to Authority* (Harper & Row, New York, 1974).

12 Y. Lacoste, *La Géographie, ça sert d'abord à faire la guerre* (Francois Maspero, Paris, 1976).

13 J. Ruskin, *Modern Painters*, ed. A. J. Finberg (Bell & Sons, London, 1937; first published 1843), vol. IV, ch. XVIII, section 5.

14 E. Cobb, *The Ecology of Imagination in Childhood* (Columbia University Press, New York, 1977). For an excellent geographical account of some of the complex links between people and environments (he phrases them in terms of 'geodiversity'), see P. T. Karjalainen, 'Geodiversity as a lived world: on the geography of existence', University of Joensuu Publications in Social Science no. 7 (1986), Joensuu, Finland.

15 B. Berenson, *Sketch for a Self-Portrait* (Pantheon, Toronto, 1949), p. 20.

16 R. Descartes, *Discourse on Method*, Discourse 6 (Penguin, Harmondsworth, 1974; first published 1637), p. 78.

17 J.-P. Sartre, *Nausea* (New Directions Paperback, New York, 1964; Penguin, Harmondsworth, 1965; first published 1938), p. 125.

18 Ruskin, *Modern Painters*, vol. III, ch. XVII, sections 4 and 5.

19 Ibid., vol. IV, ch. XVIII, section 4.

20 I. Murdoch, *The Sovereignty of Good over Other Concepts* (Cambridge University Press, Cambridge, 1967), p. 12.

Chapter fifteen

Forms of life, history, and mind: an idealist proposal for integrating perception and behaviour in human geography

Leonard Guelke

The emergence of perception studies in geography was a logical development of a line of geographic thought that began with environmental determinism. The environmental determinist thesis provided human geography with a powerful unifying theme. In hypothesizing that human society was a causal outcome of the natural environment, environmental determinists knitted physical and human geography together in a unified whole.[1] The explanation of how and why societies had developed became a function of physical factors. Environmental determinism provided a coherent unifying focus for the discipline of geography which was able within its framework to ask major questions about human society. The very essence of historical and social life was premised on its geographical base.

The causal mechanism of the environment was never directly established and seemed to run counter to much empirical evidence. As this evidence accumulated geographers introduced a number of modifications in an attempt to salvage something from the original thesis. The notions of environmental possibilism and probabilism were introduced.[2] In the former case, the environment was credited with imposing limits on human activity, but these limits did allow for considerable variations among societies inhabiting similar environments. In the latter case, one particular solution among many possible ones was deemed to have a greater probability of occurring. These modifications certainly gave scholars 'more room to manoeuvre', but the power of the environment as an explanatory factor was greatly reduced.

The notion that individuals acted in the environment in terms of how they perceived it provided another escape from environmental determinism. This idea was put forward by Sauer and Wright, but it was William Kirk who developed the idea in a detailed and logical way.[3] Kirk drew a distinction between the phenomenal environment and the behavioural environment. The behavioural environment was that part of the environment which an individual perceived and understood, and provided the basis of his or her planned actions. The Kirk model pointed the way to explaining human activity in the environment on its own terms and in ways that avoided any causal assumptions about the environment influencing human actions.

In adopting the idea that people acted in the environment as it was perceived, geographers retained an essentially positivistic view of human geography. People were, on the basis of their unique experience, equipped with 'filters' which accounted for their different reactions to the 'same' environment. The Kirk model retained the notion that there was in fact a 'real' environment independent of the human mind.[4] This view of the 'perceived' and the 'real' was implicitly adopted as the basis of much work in perception geography. Given this point of view, it was but a short step to making the task of perception geography the explanation of the 'filtered worlds' of individual perceptions of reality. This orientation led directly towards psychology, with its long record of active research in human perception. Many geographers became more concerned with perception itself than using it as a basis for understanding human activity in the environment.[5]

There were also many geographers who sought to relate behaviour and perception in seeking to understand patterns of human use and occupance of the earth.[6] In this endeavour they were at a disadvantage *vis-à-vis* the environmental determinists. In the environmental determinist thesis society was a product of environment – and geography assumed a pre-eminent role among the sciences clearly focused on major questions of human development. In its restated environmental perception form the role of the environment was less clear.[7]

There is not a clear correspondence between the way people perceive the environment and their actions, and one would not expect such a correspondence. People act in the environment to achieve a variety of social and economic goals with varying degrees of environmental awareness. If environmental ideas are to be properly integrated in explanations of human behaviour they must be understood within a broad context of social attitudes and ideas. The problem of making environmental perception studies relevant to the understanding of human activity in the environment is largely one of finding a means of situating such studies in a human geography whose subject matter and epistemological foundations have been clearly defined. With this in mind, the central idea of the thesis I wish to develop in this chapter is stated, and then defended against possible objections. My thesis is that human geography can be conceived of as the study of human forms of life, that these forms of life are historical creations of human intelligence and can most profitably be studied as expressions of mind. The reader will recognize a thesis that is clearly provocative, raising more questions than it answers when stated in this bald way. Let me begin elaborating and defending it.

Forms of life

The term 'form of life' is used here in a geographical context and is intended to convey a set or combination of elements that together give a specific pattern of living its distinctive character. The idea of a form of life implies the concrete pattern of activities that is generated by people living their lives in specific historical and social conditions and the meanings these conditions have for them. Many of these patterns of activities are associated with people making a livelihood, but the concept of 'form of life' is not restricted to economic activity, and seeks to incorporate the traditional concern of human geography to understand human activities in their physical and regional settings. Examples of forms of life might include the seasonal rhythms of nomadic pastoralists, working-class urban life in the early industrial revolution, migrant labour in colonial Africa and its impact on women cultivators, or the gentrification of urban slums and its meaning for new and old residents of such areas. However a group or theme is defined, the human geographer will seek to investigate the topic in the context of environment and physical setting. The term 'form of life' is used to denote a way of living in the physical environment in a similar sense to which the term 'form of government' might be used to describe a political system. Thus, a geographer might study the meaning and implications of industrialism or colonialism as a form of life in the way a political scientist might analyse the nature of democracy as a form of government.

The term 'form of life' as I have used it here has no connection with Wittgenstein's philosophical use of this term, but rather derives from well-established traditions of thought in geography. The geographer Vidal de la Blache and some of his followers put forward the concept of form of life or '*genre de vie*' as a central theme of human geography.[8] For Vidal, the concept of *genre de vie* was defined as a functionally unified pattern of living, which characterized certain livelihood groups such as agriculturalists and nomads.[9] Livelihood was the central concept of *genre de vie*, which sought to uncover the physical, social and psychological bonds that cemented a cultural group.[10] The concept of *genre de vie* was less concerned with the internal principles underlying culture than it was with its external manifestations. This emphasis was akin to the idea of culture adopted by Sauer and many of his students.[11] The concept of culture as a driving force in the human occupance and use of the earth fails to take account of history conceived of as a power struggle for the control of human and natural resources.[12] The point is that cultures themselves are not static, but dynamic and evolve as individuals and groups struggle to achieve some control over their lives.

The notion of a social order is essentially historical. A form of life is seen as emanating from individual and group struggles for power and is seen to be dependent on internal principles, principles which need to be analysed and understood if the form of life emanating from them is to be understood in its historical context. The way in which groups create their forms of life can be studied at many levels of generalization. At the meso level one will be concerned with the common elements characteristic of civilizations; at micro scales the underlying values of a civilization will be assumed and the power struggles defining its regional variants will be studied in farm, village, town, and other local settings.

The central point is that any penetrating analysis of a form of life must deal with ideas. Human society has created its own forms of life. A form of life is the concrete expression of the fundamental values of a people: it expresses their priorities in time and space. An understanding of forms of life, it will be argued here, must be sought in relation to the assumptions and principles of organization of the group under study. This task can be conceived of as an elucidation and intellectual commentary on how people in specific circumstances have organized their lives. Or to put this thought in a slightly different way: the human geographer is concerned to show how the principles and assumptions of a social order are manifested in the form of life of the group, which bases its existence on them.

The idea of a form of life integrates social, environmental, and technological elements. A form of life is shaped by underlying values and technological capabilities: it is the concrete manifestation of the ways in which human beings have sought to satisfy their economic and social needs under a multitude of environmental conditions. The concept encompasses a wide range of different possibilities in terms of scale and approach. It encompasses, for example, the way people have arranged themselves spatially on the earth's surface, and the implications of these arrangements for the conduct of social and economic life. The concept of a form of life can encompass such topics as the lives of pastoral nomads securing a livelihood by seasonal exploitation of different environments, the creation of slums in many modern industrial cities or the patterns of tourism created by affluence and mass air transportation.

The idea of the region has a long and distinguished history in geography and it overlaps in many ways with the concept of a form of life.[13] However, the idea of a form of life is at once narrower and broader than the idea of a region. The form-of-life concept puts the focus on human beings rather than the physical landscape as such, but includes landscape in relation to the form of life under study. The idea of a geographical region implies a bounded territory. Although geographers advocating regional study tried to escape from this implication of the concept, at least as far as the drawing of rigid lines on maps was concerned, the concept itself has the idea of a bounded area built into its essence.[14] The form-of-life concept has no such limitation, and has the flexibility to deal with human activity which is not easily classified on a regional basis. The suggestion of Hart that geographers study city regions as a way of making the regional concept more relevant to modern times can be accommodated within a form-of-life framework.[15] Cities might, under this rubric, be studied both systematically and regionally, depending upon the way people had organized their lives.

In defining a distinctive subject matter for human geography, the vexing problem of whether geography is special in the sense of being a synthesizing discipline can be resolved.[16] In the same way as geologists are concerned to study rocks, the geographer is conceived of as being concerned to study forms of life. The synthesis of elements to be studied exists in the subject matter itself. The role of the geographer is not to put elements together; it is, like that of any other scientist or scholar, to pull them apart, to discover what elements constitute a particular form of life and how it came to have its distinctive character. In looking at a specific situation it might be necessary to investigate economic, environmental, political, and social factors, but, however widely one might search, the task remains focused on how these different factors have contributed to shaping or maintaining a form of life. Of course, the analysis of how different elements contribute to the whole implies that the relation of parts to the whole are kept in mind and to this extent geography, like most other disciplines, has an integrating aspect, which might be considered to constitute a kind of synthesis. This is a far cry from envisaging geography as a synthesizing discipline with a special mandate to combine the results of other fields.

Although the concept of a form of life necessitates an analysis of spatial elements and their implications, space as such is not part of the definition. The notion that geography is spatial analysis helped scatter geographers all over the systematic disciplines without providing an integrative focus for the discipline. Geographers in the spatial tradition soon found that they had far more to say to their colleagues in the systematic fields they happened to be studying and virtually nothing to say to their colleagues in geography who were not in the same systematic field. Thus, for example, the spatial study of voting patterns provides material of interest to political scientists, but fails to provide a basis of substantive interest to other geographers. With the decline of regional geography, the discipline lacked a focus, and its continued existence as a discipline was, not surprisingly, questioned by many scholars and administrators. In defining geography as the study of human forms of life one does not preclude spatial studies or indeed systematic spatial studies, but one does provide a focus to ensure spatial studies do not become the basis of a fragmented discipline comprised of specialist studies with nothing in common.

The notion that human geographers study forms of life clearly implies a separation of human and physical geography that some geographers might find unacceptable. Yet the separation is not as decisive as it might appear. In separating human geography from physical geography one merely insists that the physical environment be evaluated from a human perspective, in terms of what it means for the human use and occupance of the earth. This perspective is quite different from that of a physical geographer who might be interested in geomorphology as a problem in applied physics and chemistry. In other words, the physical geographer concerned with understanding the mechanisms of slope formation can often do so without direct reference to human beings. Having said this, it would also be clear that the human geographer will often be able to use the results of physical geography by giving these results a human perspective and the physical geographer, concerned with human impact on the landscape, will often be able to use the results of human geographers who have studied the form of life of the people inhabiting the area under study. There is no implication that human geographers do not need physical geography or that physical geographers can get along without human geography.

The concept of a form of life cuts through the conventional classification of the social sciences. It is not sociology, although sociological elements are present and sociological factors will often play important roles in explanations. It is not economics, but economic factors may easily be integrated into the analysis. It is not history, but it forces geographers to take history seriously by emphasizing continuity and change. It is not environmental determinism, but it does take the environment seriously by conceiving of a form of life as something rooted in particular places, but recognizes that the meaning of environment is a function of human intentions and knowledge. Above all, the idea of a form of life emphasizes the need for a coherent human geography which incorporates the regional, spatial, and humankind–environment inter-action traditions of the discipline. Although it provides some restriction on what one might consider geography, the concept is open-ended in the sense that the ways in which forms of life might be studied are limited only by the imagination of geographers. It is also permissive in terms of epistemological orientations. Positivists, existentialists, phenomenologists, realists, pragmatists, and Marxists might all have something to offer.

Philosophical foundations

The term idealism is extremely broad and identifies a tradition in Western philosophy that might be better expressed in the word 'ideaism'. The central importance of ideas, spirit, or reason in the human saga is recognized in idealist philosophies, which have included among its proponents such intellectual giants as Plato, Berkeley, Kant, and Hegel. The importance of idealism as a fundamental epistemology in geography rests on some basic assumptions. The first assumption is that reality is a product of mind. Without wishing to become unduly abstruse, this principle asserts that the real world, or what we think of as the real world, is a mental construct. The interpretation of the phenomena we experience and perceive depends entirely on the ideas and beliefs of the perceiver. Without ideas to organize our sensations nothing can exist but an undifferentiated jumble of sensations without order or meaning.

If the proposition in the preceding paragraph is accepted, it follows that one's ideas become the basis of the reality one constructs. Ideas provide the organizing principles on which the world is made understandable, and understanding is in turn the basis for action. There can be no reality except that reality which mind has constructed. Here it might be objected, however, that surely there is a reality beyond the mind: the reality of earthquakes, crops in fields, snow-capped mountains. These phenomena are really there whether we wish to acknowledge them or not. The idealist position is that such realities as I have just listed are not 'real' in any sense until they have been perceived, classified, and interpreted by mind.

The idealist approach seeks to build upon the work of such scholars as Wright, Kirk, Lowenthal, Brookfield, and Prince, all of whom sought to deal with the subjectivity inherent in human perception in a cultural-historical context.[17] These efforts at redirecting geographical research away from materialism met with limited success, and when environmental perception studies became an important theme of modern geography they drew their inspiration from positivistic social science rather than humanistic philosophies of meaning.[18] The idealist philosophy provides a basis for integrating perception studies within a broader historical and intellectual context. The concept of world-view – for individuals and societies – permits a scholar to focus on the intellectual assumptions through which the world is interpreted. The task of the human geographer is seen to involve relating human behaviour to the total cultural context in which it is embedded. The importance of environmental ideas is not underestimated, but it is also recognized that activity in the environment is a function of a constellation of ideas relating an individual or group to society and environment.

The nature of 'reality' has changed as a function of changing ideas. The medieval European world was based on ideas and principles of order that are quite different from those of the modern world. Yet we cannot claim that modern science has produced a 'real' picture of the world, because science is not static. As knowledge advances, ideas about the nature of reality change. There no more exists a real world today than there existed a real world in medieval times. In both cases what passes or passed for 'reality' is merely a way of referring to an agreed-upon world constructed on the basis of shared assumptions, which of necessity are historically specific.

The critical component of an historical sequence of events is thought or reason, a position first advocated in a systematic way by the philosopher Hegel. For Hegel history is underpinned by philosophy: 'The sole thought which philosophy brings to the treatment of history is the simple concept of *Reason*: that Reason is the law of the world and that, therefore, in world history, things have come about rationally.'[19] Many modern critics of this position have pointed to its lack of concern with human psychology, and economic and environmental factors, but these criticisms miss their mark. Hegel, of course, was a product of his time, and his own history fully reflects the biases and preoccupations of contemporary scholarship. The central tenet of his philosophy of history, however, is not affected by such limitation in his own era. Hegel was putting a case for reason as the foundation of history and historical change.

Although reason underlies human societies and defines their natures, this is not to say that reason exists as a force independent and divorced from human psychological and material existence. Rather, reason is fully integrated in the biological facts of human life and, indeed, it owes its power to them. The importance of reason lies in its potential for self-development. Ideas grow on the basis of existing ideas. Human intellectual development is an historical phenomenon involving the growth and change of ideas.[20] These changes are not random, but logically entailed by developing new implications of existing ideas. As human life created the rules of its own existence, it evolved methods of communication, especially languages, and through them the basis of intellectual communication was established.

A form of life is the historical creation of mind.[21] This is the foundation on which this idealist view of human geography rests. This position incorporates elements that can be found in other approaches. It claims distinctiveness in the way its three central elements (form of life, history and mind) are blended together.

The mind creates history and was created historically. In other words, the basis of reality is mind which interprets the world. This point is fundamental; but why do geographers need to be concerned with 'mind', since they are not intellectual historians? The geographer's interest in mind stems from his or her interest in the way people live and have settled the earth. These activities are produced by people thinking independently and making decisions as members of social groups. The shared assumptions and ideas create the world, the real world. Geographers need to examine these assumptions and ideas in order to relate thought and action and to understand how people behave in their worlds.

In interpreting the world, geographers must show how reason has created a variety of forms of life which are in constant flux. Although the existence of 'reality' is not denied by idealists, it is not forces external to the human body which control development. It is human interpretation and assessment of the external world which decides how people will respond to it. Here we must be careful, however. Too often geographers have sought to interpret behaviour in the environment as a function of environmental ideas.[22] The important point is that action be linked to the total intellectual context in which it occurs. If change occurs, the task of the geographer is to show how the group developed its new ideas logically out of its old ones. Geographical change is like historical change in that it involves the self-development of reason. Idealist understanding has little to do with psychology: the essence of human action is plan and design.

In geography a balance between life form, mind, and history is needed.[23] Geography or human geography is a discipline concerned mainly with ordinary life, with people in places. The scholar's task is to make sense of how people live and have lived in a way that does not lead ever backwards or ever outwards in search of explanations. There are too many other scholars far better qualified than geographers to investigate the world system and the causal mechanisms of its economy. However, geographers can deal with people, ideas, and place in ways that provide connections and open questions. The focus of geographers on people and place is not a new idea: the proposal is new only to the extent that it suggests ways in which people and places can be studied in a fully intellectual and historical context.

Idealism, in emphasizing mind and ideas as the foundation of forms of life, opens up a vast subject matter for intellectual exploration. In positivistic studies the people act in ways that appear inevitable: the material reality, as it were, dominates explanations to the point at which everything that exists or has existed has an inevitable look to it.[24] This inevitability is implicit in the use of the word 'process' to describe how systems work. The idealist emphasizes human control of the events creating the human reality. Why were these decisions made by this people at that time? Why not earlier or later? Why not other decisions from the ones adopted? What do the ways in which people use their environments tell one about their fundamental values?

The human geographer's involvement with history is quite different from that of the political historian who is concerned to analyse political power struggles in specific situations. The human geographer is concerned with the results of these power struggles in so far as they provide a foundation for everyday life. An historian, for example, might analyse the origin of liberalism in the United States. The geographer's concern would be in analysing how this principle manifested itself in concrete economic and social circumstances. What is the meaning of freedom on the ground for different peoples in different circumstances? How have fundamental ideas affected the human use and occupance of the earth? In other words, the idealist problematizes much that might be taken for granted in positivistic geography. The primary goal is to analyse the experience of life in a way that makes possible an intellectual analysis, which is not ideologically committed, and which takes the historical foundations of the world's cultures and societies seriously.

Forms of life are created and destroyed as principles of organization and fundamental assumptions evolve and change. Human geography is concerned to understand and explicate the ways in which ideas have manifested themselves in specific forms of life. Thus, for example, the idea of slavery created a variety of distinctive forms of life, which owed their existence to an acceptance of this principle. When slavery was abolished in various parts of the world new forms of life emerged based on new principles of social and economic organization and new ways of labour exploitation.

Although human geographers by the nature of their subject matter are not concerned with intellectual history as such, the analysis of forms of life should be conceived of as an essentially intellectual task. In making sense of how a social order functions, the geographer is concerned to show how ideas provide the basis of organization and meaning in a given geographical context. This point is nicely demonstrated in Tannenbaum's classic study *Slave and Citizen.*[25] In this work Tannenbaum argued that the experience of slavery in the United States was particularly harsh, because slavery could only be justified on the basis that slaves were property. The liberal ideology of 'all men are created equal' meant that a slave was either property or a full citizen. In contrast, in Brazil the Roman Catholic ideology of society based on concepts of hierarchical organization and levels or stations of social status enabled the slave to be incorporated at the base of a social system in which individuals enjoyed a variety of rights dependent on rank. The important point is that slaves in Brazil were seen as part of society, not separate from it, and their humanity was not questioned as it was in the US South.

The above example is intended to show the potential power of an analysis in which fundamental ideas are taken seriously. This kind of analysis does not give a scholar licence to speculate. Interpretations need to be supported with appropriate evidence; indeed, a major task of analysis is to point out the kinds of evidence that might be pertinent to a specific interpretation. Thus, Tannenbaum's work generated a considerable amount of empirical research by scholars searching for ways to test his thesis. This process did not leave Tannenbaum's thesis unscathed, but it most certainly enlarged our understanding of the nature of slavery and slave societies.[26]

There are an infinite number of potential topics open to geographical investigation. Forms of life can be approached from both regional and systematic perspectives. In studies in the United States one might define a group on the basis of race or ethnicity in terms of status, or wealth, or gender, or in terms of geographical regions ranging from a city block to entire states. A group's way of life, however one wishes to define it, can be investigated in terms of ideas and assumptions of the study group as well as the society at large. A theme in the geography of the United States might well be liberty. This fundamental principle needs to be defined and its meaning for different groups at different times and places needs to be explicated. Thus, for example, the nature of poverty, resource exploitation, and land-use planning in the United States cannot be understood unless one also understands how Americans have defined freedom. Such fundamental ideas as individualism, freedom, and democracy are reflected in the forms of life of Americans, and a task of geographers is to connect the ideal and reality in the intellectual analysis of their forms of life.

The adoption of an idealist view of human geography provides a basis for a coherent discipline. The ideas and assumptions underlying particular life forms allow the human geographer to connect a particular case study with a larger historical context. Indeed, the purpose of case studies would be to illuminate how the specific experience of a group or community exemplifies the ideas which gave it (the experience) meaning and how these ideas relate to the larger picture. The focus on ideas provides an alternative to the theories of the positivist, without geographers having to revert to description. The whole point of the approach lies in its recognition that ideas are dynamic, that change is normal, and that the logic of historical change is human reason developing itself in concrete historical circumstances. The geographers' task is seen as the explication of the implications of change for the way people live and experience life.

An idealist human geography does not lend itself to a formula approach as exemplified in the scientific method. Any situation can be illuminated in a variety of ways: the level and power of an analysis will largely depend on the backgrounds and imagination of geographers themselves and the existence of a strong critical environment.[27] It is also recognized that scholars will be prisoners of their time in terms of their ability to recognize and analyse certain phenomena.[28] In our times race and gender have been redefined in ways that have broadened our perspectives of society and made studies possible where few existed before. The idealist perspective recognizes that there are limits to objectivity, but also emphasizes the need for evidence in the quest for understanding.[29]

Comparison with other approaches

Although one might be hard put to find geographers who still advocate an environmental determinist thesis, this position needs to be scrutinized because it illuminates some important philosophical points. The environmental determinists thought that they had found the fundamental cause of differences in human forms of life in the natural environment. The approach was intellectually exciting in that it was able to reduce complex situations to the operation of a clearly designated set of causal variables. It was also challenging. Geographers had as their research agenda the daunting task of showing how individual case-studies fitted within the general thesis. And finally, the approach provided geographers with a clearly demarcated subject matter and set of problems in the humankind–environment tradition. The principal shortcoming of environmental determinism was that the causal assumption built into its definition was empirically unverifiable.

In spite of its philosophical weakness some excellent work was done by environmentalist geographers. Ellen Churchill Semple's *The Geography of the Mediterranean Region* is an example of such a work.[30] The fact that environmental determinists were concerned with questions of human ways of life and civilization accounts for the intellectual strength of their best work. This strength is retained in defining geography as the study of forms of life. The causal thesis of environmental determinism is abandoned and replaced with an open-ended commitment to understanding society in terms of ideas. However, in emphasizing ideas one does not abandon the consideration of the environment but looks at what people have done with it and to it in relation to their historically specific perceptions of it. The intellectual task becomes one of writing a critical commentary on human forms of life as these related to the environment in the context of historical change.

When geographers abandoned the environmentalist thesis there was not much to put in its place. What had made environmental determinism intellectually exciting made much of the cultural-regional geography of the Sauer–Hartshorne era overly descriptive. Geographers continued to study human occupancy and use of the earth, but without any clearly articulated intellectual goal. The application of an empirical methodology generated many descriptive studies. At its best, the geography of the 1930s, 1940s, and 1950s was not without merit, and certainly geographers often made more of their studies than these descriptive methods; but when they did good work they did it in spite of rather than because of this methodology.

The critical weakness of regional geography was not its lack of theory, but rather its lack of an epistemology which fostered the integration of specific case-studies within broader contexts. Idealism seeks to address this weakness. The assumption that human societies are founded on ideas provides a way of relating individual case-studies to each other and developing broadly based interpretative themes. How have the principles on which societies have been based been reflected in their various forms of life? The geography of colonial rule, for example, might provide a basis for a general historical geography or one for several case-studies in which specific features of colonial rule in regional contexts were addressed. The point about case-studies is that they would not stand alone, but would be linked to broader issues. One might, for example, show how colonial administrators dealt with problems – as they defined them – of resource development. Although the study might involve a specific situation, the case-study would be intellectually valuable in so far as it illuminated the operation of the idea of colonialism.

The rise of quantitative and theoretical geography was a reaction to the descriptive character of much regional geography. Yet it was not essentially a revolution, but rather the more rigorous application of scientific methods, already well accepted, to geographical problems.[31] It sharpened and strengthened the analysis of location and linked geography to other social sciences.[32] The new geography, in emphasizing space as a fundamental concept, failed to provide a means of integrating individual case studies within a broader disciplinary perspective. The explanatory schema were largely mechanistic and ahistorical. In a quest for generality the concern with the specific or unique characters of regions and places was largely abandoned. The lack of a holistic reference point, the general absence of an historical perspective, and a preoccupation with what could be measured meant that many fundamental questions relating to values in society were never raised.[33] These weaknesses fostered a search for new epistemologies by many geographers who wanted to strengthen the intellectual foundations of the discipline.

The need to instil a stronger explanatory content in geography provided a stimulus to the growth of Marxist ideas. The Marxists provided a much-needed historical perspective to geographical work, albeit within a framework of a dialectical materialism. Idealists and Marxists share a common assumption in according history a central role in explanation and understanding. They differ as to the motor of change. The Marxist sees material conditions as fundamental, the idealist gives precedence to ideas.

For Marxists, history is governed by inexorable materialist forces, although human agency is not entirely precluded. In adopting a strong materialist view of change Marxists put themselves in the same company as environmental determinists. Although they differ on the factors that shape society, they both agree that those factors are material ones and that the task of scholarship is to interpret specific situations in relation to underlying material forces which control them.

The idealist approach is open-ended in that it does not identify historical or social change with a specific cause or sets of causes acting on society. Human geographers in the positivistic tradition of social science have frequently resorted to theories and models derived from such areas as classical economics which have built-in value assumptions. Marxists have criticized such theories and models as implicitly incorporating a political position favouring a continuation of the *status quo*. Yet Marxists have adopted an alternative perspective which includes their own assumptions about the nature of historical process. Both approaches assume that history is a process governed by laws of human behaviour. The idealist, in emphasizing the importance of reason as it develops in concrete situations, is able to analyse power relations without general presuppositions about ideal or expected outcomes. History is treated as open-ended. These two perspectives are fundamentally different. In the words of Lasch:

> Anyone who insists on the historical importance of human actions, and who sees history not as an abstract social 'process' but as the product of concrete struggles for power, finds himself at odds with the main tradition of the social sciences, which affirms the contrary principle that society runs according to laws of its own.[34]

From an idealist perspective, historical materialism exhibits a fundamental weakness by relegating ideas to a secondary role. Historical materialism has many interpretations and, in some of its variants and derivatives, ideas are accorded a major role and individual agency is recognized as of fundamental importance.[35] In seeking to provide a more realistic interpretation of the historical-materialist thesis, geographers and other scholars have come to straddle a half-way position between idealism and materialism. The position provides plenty of flexibility, and ensures that interpretations of individual cases can be fitted within the materialist thesis.[36] For the idealist, a stronger philosophical position is achieved simply by lopping off the materialist portion of the thesis. In doing this the importance of human agency is fully recognized and the need to reconcile individual actions with general causal forces is removed.

All individuals are located in specific historical contexts, and how and what they think about will inevitably be shaped by their historical situations. What are objective structures for the Marxist are for the idealist shared assumptions and ideas. The end result in so far as explanation goes might not in some cases be that different, but the final positions are very far apart. The idealist sees history as an open-ended battle of ideas, in which individuals may seek to achieve a variety of goals. Whatever individuals do struggle to achieve, their weapons are ideas and these weapons are in fact the real stuff of historical change. This open-ended approach which does not commit itself to theory contrasts with the neo-Marxist position which, when all the qualifications have been stripped away, remains a philosophy in which material forces rather than forces of the spirit (ideas) are considered to control the 'processes' of history.

If the emergence of Marxist geography was one reaction to the dominance of positivistic social science in the geography of the 1960s, the other was the growth of humanistic approaches in the discipline. Humanistic geographers drew their inspiration from several philosophical and intellectual sources among which works by Cassirer, Husserl, Heidegger, Merleau-Ponty, Langer, and Sartre figured prominently.[37] In addition, a variety of humanistic positions were developed on the basis of individual philosophers. Wittgenstein, Buber, Dilthey, and James and Dewey all provided the foundation ideas for specific humanistic proposals.[38] If these individual humanistic proposals varied widely, they all shared with idealism a concern with the subjective dimensions of human experience.

Although there were similarities, the questions of concern to most humanistic geographers were rather different from those of the idealist. An interest in the individual experience of the world in terms of a radical description of that experience was a fundamental theme of humanistic geography. The perspective has emphasized introspection and the symbolic, aesthetic, and ethical meaning of the life experience.[39] As an intellectual enterprise, humanistic geography did not develop a strong historical dimension and the exploration of the subjective realm of human consciousness often depended on the individual rather than on external sources for verification.[40] In these two areas, namely history and evidence, the idealist project has a very different orientation to most humanistic geography.

Humanistic geographers have not had a clear idea of the kind of distinctive contribution geographers might make to the intellectual world. The exploration of individual consciousness has assumed a variety of forms and taken geographers off in many directions. The idea of human forms of life is in keeping with much humanistic geography, and could give humanistic studies a sharper geographical focus by providing limits to the sorts of questions geographers should ask as geographers. The fundamental point, however, is one of intellectual coherence. Idealism in emphasizing the historical context of mind provides a basis on which individual case-studies can be linked to each other and imbued with wider meaning.

Conclusion

The concept of form of life seeks to meet the minimal requirement of subject matter in human geography. It is in the traditions of human geography and incorporates regional themes typified in the idea of place and landscape. Yet the concept is also able to accommodate a variety of systematic themes involving space and physical environment. The concept supports enquiries at a range of scales from local to global. Yet it is not an empty definition. In insisting that geographers relate their work to forms of life, the concept rules out many kinds of purely spatial analysis from geography as well as esoteric phenomenological studies unrelated to human forms of life. The goal of identifying a specifically geographical subject matter establishes an integrating theme for work by geographers and provides a basis on which human perception and behaviour in the environment can be analysed and understood.

When geographers abandoned environmental determinism, human geography was left without an overriding intellectual task. Neither descriptive regionalism, quantitative and theoretical spatial studies, Marxist-inspired analysis of social and economic geography, nor an open-ended phenomenology was able to provide a distinctive role for geography as an intellectually credible and coherent discipline.[41] A radical reversal of the claim that the natural environment determines society is that society makes itself and creates its own environments. But it does not do this instantaneously. Each society is the outcome of its own unique history, which has provided its underlying assumptions of order and defined the way people live in the environment. The historical legacy is most importantly a legacy of ideas.

The motor of historical change is not the environment, the economy, or any other force, but human reason. Ideas have the unique capacity of logical development and humans have changed their forms of life and their relationships to the environment in an orderly and logical way. In accepting this fundamental point the thesis of environmental determinism is, as it were, stood on its head, and the overriding task of human geography becomes the investigation of the logical relationships among the ideas contained in human forms of life in the physical environment.

The physical environment acquires new intellectual meaning in the context of historical forms of life. The ways different peoples have interpreted their relationships to the environment and the ways in which these relationships have been connected to their specific visions of life and society become matters of central concern to human geography. An idealist analysis seeks to make sense of human patterns of living by asking questions like: 'What did these people think they were doing when they took this or that action?' or 'What was this group's understanding of its own situation and on what assumptions was it based?'

The idea that the environment must be understood in terms of the way people perceived it was a major step forward in human geography. It recognized that it was people's understanding of environment, not environment itself, that shaped their actions, but it did not elaborate on a means of determining the meaning of the environment itself in the broader context of a society's priorities and visions of order. A human geography dedicated to investigating human activities or forms of life in their historical contexts seeks to establish a broader basis for understanding the character of human activities on the surface of the earth and their connections to the physical environment.

Acknowledgement

I am grateful to my colleagues Jean Andrey, Roy Officer, and Peter Nash for their critical comments on an earlier draft of this chapter. The responsibility for the ideas expressed herein remains entirely mine.

Notes

1 This integration of physical and human geography is typified in the work of such environmental determinists as Ellsworth Huntington, Griffith Taylor, and Ellen Churchill Semple.
2 G. Tatham, 'Environmentalism and possibilism', in G. Taylor (ed.), *Geography in the Twentieth Century*, 3rd edn (Methuen, London, 1953), pp. 128–62.

306

3 C. O. Sauer, 'The morphology of landscape', University of California
   Publications in Geography no. 2 (1925), University of California,
   Berkeley, pp. 19–54; J. K. Wright, '*Terrae incognitae*: the place of the
   imagination in geography', *Annals of the Association of American
   Geographers*, vol. 37 (1947), pp. 1–15; W. Kirk, 'Historical geography
   and the concept of the behavioural environment', *Indian Geographical
   Journal*, Silver Jubilee Volume (1952), pp. 152–60.

4 Kirk, 'Historical geography and the concept of the behavioural
   environment', p. 159.

5 For example, T. F. Saarinen, 'Perception of the drought hazard on the
   Great Plains', Research Paper no. 106 (1966), Department of
   Geography, University of Chicago; and P. Gould and R. White,
   Mental Maps (Penguin, Harmondsworth, 1974).

6 Most notably, the studies of human adjustment to natural hazards
   emanating from the seminal work of Gilbert White at the University of
   Chicago in the 1940s, 1950s, and 1960s.

7 L. Guelke, 'Interdisciplinary research and environmental perception',
   Proceedings of the Association of American Geographers, vol. 5
   (1976), pp. 184–8.

8 A. Buttimer, 'Society and milieu in the French geographic tradition',
   Monograph no. 6 (1971), Association of American Geographers,
   Washington, DC, pp. 52–7. In acknowledging that the source of
   inspiration for the concept 'forms of life' was the *genre de vie* concept
   of French human geography, I do not imply that the two concepts are
   equivalent. The idea of a form of life as I have presented it here must
   be judged on its merits and in relation to the arguments set out in
   support of it, many of which would not have been approved of and
   were not part of the concept *genre de vie*. In addition, the form-of-life
   concept seeks to establish a basis for human geography as a coherent
   discipline on its own and is not conceived of as one element in a cluster
   of elements defining human geography.

9 Ibid., p. 51.

10 M. Sorre, 'The concept of *genre de vie*', in P. L. Wagner and M. W.
    Mikesell (eds), *Readings in Cultural Geography* (University of
    Chicago Press, Chicago, 1961).

11 C. O. Sauer, 'Foreword to historical geography', *Annals of the
    Association of American Geographers*, vol. 31 (1941), pp. 1–24.

12 J. S. Duncan, 'The superorganic in American cultural geography',
    Annals of the Association of American Geographers, vol. 70 (1980),
    pp. 181–98.

13 J. F. Hart, 'The highest form of the geographer's art', *Annals of the
    Association of American Geographers*, vol. 72 (1982), pp. 1–29.

14 D. Whittlesey, 'The regional concept and the regional method', in P.
    E. James and C. F. Jones (eds), *American Geography: Inventory and
    Prospect* (Syracuse University Press, Syracuse, NY, 1954), pp. 21–68.

15 Hart, 'The highest form of the geographer's art', p. 24.

16 R. C. Harris, 'Theory and synthesis in historical geography', *Canadian
    Geographer*, vol. 15 (1971), pp. 157–72.

17 Wright, *Terrae incognitae*; Kirk, 'Historical geography and the
    concept of the behavioural environment'; D. Lowenthal, 'Geography,
    experience, and imagination: towards a geographical epistemology',
    Annals of the Association of American Geographers, vol. 51 (1961),
    pp. 241–60; H. C. Brookfield, 'On the environment as perceived',
    Progress in Geography, vol. 1 (1969), pp. 51–80; H. Prince, 'Real,
    imagined and abstract worlds of the past', *Progress in Geography*,
    vol. 3 (1971), pp. 1–86.
18 T. E. Bunting and L. Guelke, 'Behavioral and perception geography:
    a critical appraisal', *Annals of the Association of American
    Geographers*, vol. 69 (1979), pp. 448–62; D. Ley, 'Behavioral
    geography and the philosophies of meaning', in K. R. Cox and R. G.
    Golledge (eds), *Behavioral Problems in Geography Revisited*
    (Methuen, New York and London, 1981), pp. 209–30.
19 G. W. F. Hegel, *Reason in History*, translated with an introduction by
    R. S. Hartman (Liberal Arts Press, New York, 1953).
20 This position is developed in R. G. Collingwood, *The Idea of History*
    (Oxford University Press, New York, 1956; first published 1946).
    Although Collingwood clearly based his philosophy of history
    on his predecessors, notably Kant, Vico, and Hegel, he was an
    original thinker and sought to go beyond these philosophers in
    developing a coherent and logically defensible concept of history. For
    a discussion of Collingwood's philosophy of history in relation to Kant
    see D. N. Livingstone and R. T. Harrison, 'Immanuel Kant,
    subjectivism, and human geography: a preliminary investigation',
    Transactions, Institute of British Geographers, New Series, vol. 6
    (1981), pp. 359–74.
21 The revival of interest in Vico might provide a stimulus for
    geographers to explore the proposition: see W. J. Mills, 'Positivism
    reversed: the relevance of Giambattista Vico', *Transactions, Institute
    of British Geographers*, New Series, vol. 7 (1982), pp. 1–14. Indeed,
    the ideas of Vico and his *verum ipsum factum* principle provide a point
    of departure for the idealist notion that as humans we are in a unique
    position to understand history because we ourselves are its creators.
22 Bunting and Guelke, 'Behavioral and perception geography', p. 460.
23 L. Guelke, *Historical Understanding in Geography: An Idealist
    Approach* (Cambridge University Press, Cambridge, 1982).
24 D. Ley, 'Social geography and the taken-for-granted world',
    Transactions, Institute of British Geographers, New Series, vol. 2
    (1977), pp. 498–512.
25 F. Tannenbaum, *Slave and Citizen: The Negro in the Americas*
    (Knopf, New York, 1947).
26 D. B. Davis, *The Problem of Slavery in Western Culture* (Cornell
    University Press, Ithaca, NY, 1966), pp. 223–61.
27 R. C. Harris, 'The historical mind and the practice of geography', in
    D. Ley and M. S. Samuels (eds), *Humanistic Geography: Prospects
    and Problems* (Croom Helm, London; Maaroufa, Chicago, 1978),
    pp. 123–37.

28 R. Szymanski and J. W. Agnew, *Order and Skepticism: Human Geography and the Dialectic of Science* (Association of American Geographers, Washington, DC, 1981).
29 Guelke, *Historical Understanding in Geography*, pp. 49–53.
30 E. C. Semple, *The Geography of the Mediterranean Region* (AMS, New York, 1931).
31 L. Guelke, 'Geography and logical positivism', in D. T. Herbert and R. J. Johnston (eds), *Geography and the Urban Environment: Progress in Research and Applications* (John Wiley, London, 1978), p. 380.
32 D. Harvey, *Explanation in Geography* (Edward Arnold, London; St Martin's, New York, 1969).
33 H. Couclelis, 'Some second thoughts about theory in the social sciences', *Geographical Analysis*, vol. 15 (1983), pp. 28–33.
34 C. Lasch, *Haven in a Heartless World* (Basic Books, New York, 1977).
35 A. Giddens, *New Rules of Sociological Method* (Hutchinson, London, 1976); D. Gregory, 'Human agency and human geography', *Transactions, Institute of British Geographers*, New Series, vol. 6 (1981), pp. 1–18; A. Pred, 'Place as historically contingent process: structuration and the time-geography of becoming places', *Annals of the Association of American Geographers*, vol. 74 (1984), pp. 279–97; E. P. Thompson, *The Poverty of Theory and Other Essays* (Merlin Press, London, 1978).
36 For a critique of the tensions inherent in the structural Marxist position see the article by J. Duncan and D. Ley, 'Structural Marxism and human geography: a critical assessment', *Annals of the Association of American Geographers*, vol. 72 (1982), pp. 30–59; and the commentary and reply on it, V. Chouinard and R. Fincher, 'A critique of "Structural Marxism and human geography" ', *Annals of the Association of American Geographers*, vol. 73 (1983), pp. 137–46; J. Duncan and D. Ley, 'Comment in reply', *Annals of the Association of American Geographers*, vol. 73 (1983), pp. 146–50.
37 A. Buttimer, 'Values in geography', Commission on College Geography Resource Paper no. 24 (1974), Association of American Geographers, Washington, DC; A. Buttimer, 'Grasping the dynamism of lifeworld', *Annals of the Association of American Geographers*, vol. 66 (1976), pp. 277–92; J. N. Entrikin, 'Contemporary humanism in geography', *Annals of the Association of American Geographers*, vol. 66 (1976), pp. 615–32; D. Ley, 'Social geography and social action', in Ley and Samuels (eds), *Humanistic Geography*, pp. 41–57; D. C. Mercer and J. M. Powell, *Phenomenology and Related Non-Positivistic Viewpoints in the Social Sciences* (Department of Geography, Monash University, Melbourne, 1972); E. Relph, 'Phenomenology', in M. E. Harvey and B. P. Holly (eds), *Themes in Geographic Thought* (Croom Helm, London, 1981), pp. 99–114; M. S. Samuels, 'Existentialism and human geography', in Ley and Samuels (eds), *Humanistic Geography*, pp. 22–40; M. S. Samuels, 'An existential geography', in Harvey and Holly (eds),

Themes in Geographic Thought, pp. 115–32; Yi-Fu Tuan, 'Space and place: humanistic perspective', *Progress in Geography*, vol. 6 (1974), pp. 211–52; Yi-Fu Tuan, 'Humanistic geography', *Annals of the Association of American Geographers*, vol. 66 (1976), pp. 266–76.

38 M. Curry, 'Forms of life: a Wittgensteinian view', unpublished MA thesis, University of Minnesota, Minneapolis, 1980; M. Curry, 'The idealist dispute in Anglo-American geography', *Canadian Geographer,* vol. 26 (1982), pp. 37–50; M. Curry, 'The idealist dispute in Anglo-American geography: a reply', *Canadian Geographer,* vol. 26 (1982), pp. 57–9; L. Guelke, 'The idealist dispute in Anglo-American geography', *Canadian Geographer,* vol. 26 (1982), pp. 51–7; S. Hasson, 'Humanistic geography from the perspective of Martin Buber's philosophy', *Professional Geographer,* vol. 36 (1984), pp. 11–18; C. Rose, 'Human geography as text interpretation', in A. Buttimer and D. Seamon (eds), *The Human Experience of Space and Place* (St Martin's, New York, 1980), pp. 123–34; S. J. Smith, 'Practicing humanistic geography', *Annals of the Association of American Geographers*, vol. 74 (1984), pp. 353–74.

39 For example, E. Relph, 'Humanism, phenomenology and geography', *Annals of the Association of American Geographers*, vol. 67 (1977), pp. 177–9; E. Relph, *Rational Landscapes and Humanistic Geography* (Croom Helm, London, 1981); D. A. Seamon, *A Geography of the Lifeworld* (Croom Helm, London; St Martin's, New York, 1979); Yi-Fu Tuan, *Space and Place: The Perspective of Experience* (University of Minnesota Press, Minneapolis, 1977).

40 This characteristic of much humanistic geography is not by any means a necessary implication of its philosophical foundations. An example of a humanistic study incorporating large amounts of external evidence is J. Western, *Outcast Cape Town* (Allen & Unwin, London, 1981).

41 It was largely because geographers were unable to establish the intellectual value of their discipline that geography was dropped at Harvard University. See P. Nash, 'The making of a humanist geographer: a circuitous journey', in L. Guelke (ed.), 'Geography and humanistic knowledge', Department of Geography Publications no. 25 (1986), University of Waterloo, pp. 1–22.

Chapter sixteen

The behavioural environment: how, what for, and whose?

Harold Brookfield

Bill Kirk introduced the notion of the behavioural environment in a one-off special publication of the Indian Geographical Society, in which my own first-written though not first-published paper also appeared. Bill's, resurrected from obscurity into a more prominent place in 1963, has become famous, whereas mine perished in well-deserved oblivion.[1] In sketching relationships of what was then called 'urban morphology' to the terrain, however, I was, in fact, concerned with an aspect of the behavioural environment, and much of my later empirical work in Africa, the Pacific, and Malaysia has also been concerned with interpretation of the expression of human behaviour on the varied surface of the land. Now, over 30 years after I first became involved in this sort of work, is a good time to inspect it in the context of Kirk's insight and of a review and think-piece about 'the environment as perceived' which I wrote in midstream, in the early days of the modern behavioural movement in geography.[2]

Since that time perception studies have burgeoned. A whole subfield described as 'behavioural geography' emerged and has acquired a substantial literature. It incorporated the early writings of Wright, Kirk, Lowenthal, Tuan, and others, together with the early work of the natural hazards group, as its foundation literature. In what follows I shall argue that the emergence of a defined subfield around perception and behaviour in the 1970s was a mistake, something of a cop-out among geographers who sought a smoother passage among the choppy waters generated by the radicals and their conflict with the positivists. Behaviouralism, interpreted differently than it in fact was, could and should have had a major role to play in this debate, and might have done something to avoid the fission of the subject, and to wean at least some of those who followed the radical path from what has become a cul-de-sac in the hands of its devoted protagonists. However, even that depended on resolution of a methodological problem.

The first part of the chapter addresses this argument. I then go on to suggest that among the more important reasons for failure of the behavioural movement has been its rather close link with cognition – that is in its emphasis on perception and hence on the 'behavioural environment'. The proper core of behaviouralism should lie in the study of decision-making, whether in environment-use or space-use matters. A consequence is a long and very dangerous delay in reformulating both the nature–society and spatial-pattern paradigms on which the discipline of human geography rests.

Finally, I shall suggest that these failures have in part arisen from the limitations of the cognitive method. The behavioural environment of action is unquestionably important, but it contains many hands that are hidden from the actor, and is of the moment, not the long term. Moreover, it is not only very difficult to generalize about the perceived environment of individuals, but also very difficult to measure it. However we set about that task it is almost inevitable that the preconceptions of the measurer will intrude. To say this is not to say that it is wrong to try to characterize the ways in which people think: most of social science would cease to exist if this were not done. It is, however, to say that cognitive method is always selective, and time-dependent. There are two behavioural environments, that of the people being studied and that of the researcher and interpreter. It is important to recognize these facts, and not delude ourselves that we have discovered absolute truth.

What was, and what might have been

'Behavioural geography'

When behavioural geography first emerged it was in fact very much concerned with decision-making, for it was the introduction of risk and uncertainty, and of the notion of imperfect rationality, that set the ball rolling towards stochastic rather than deterministic modelling in the 1960s, and led to important statements such as that of Pred,[3] and to Harvey's attempts to incorporate perception and uncertainty into geographical explanation.[4] In the natural hazard field, the role of perception was seen as being to explain the acceptance of natural risk in terms of awareness and response.[5] Also present from the outset, however, were studies which sought directly to measure or to characterize perception of landscape and places, as well as of events.[6] Gold and Goodey argue that the early hope was to contribute substantially to public policy, especially in urban planning and resource allocation, by demonstrating the need to consider the perceptions of the people for whom planning and design were – supposedly – undertaken.[7] Already by the early 1970s, however, Walmsley could say that 'within the study of spatial behaviour, human geographers have tended to concentrate on disparate aspects so that no general approach has been in evidence'.[8] He noted an already-strong movement towards all-out phenomenology,[9] but rightly saw limited scope in this direction for the discipline as a whole. In the conflict that quickly developed between a reformist positivism and an increasingly Marxist radicalism, the weak thrust of behaviourism made little headway. Its lack of social relevance in a period when such relevance had become the central issue in human geography inevitably consigned behavioural geography to a fringing role, where it received some sharp radical criticism both as a whole[10] and more specifically against its most productive single stream, that of natural hazard study.[11]

That geographers' *Readers' Digest*, the journal *Progress in Human Geography*, is by no means an unfailing indicator of trends in the discipline, but its series of 'progress reports' are useful in this respect – despite the error of systematically separating most work on non-Western countries into a separate category. 'Behavioural geography' was not among the initial topics selected for progress reports in 1977, but from 1979 to 1985 every volume carried a report first on 'environmental perception',[12] then on 'behavioural and perceptual geography'.[13] There has also been a parallel and to some degree intersecting series on 'cultural' or 'cultural/humanistic' geography, which continues through 1988. Lowenthal linked his report to historical reconstruction. Saarinen and Sell always found a great deal of literature on which to report, stressed and rejoiced in the strong swing from positivism towards humanism, and focused in particular on the emergence of 'sense of place' as a focal area for behavioural research. By contrast, Gold and Goodey were consistently and increasingly pessimistic around a smaller selection of literature. They dealt at length with criticism, and also reported a growing schism between those who still sought to introduce behaviourism into positivistic models and the humanists. They also provided evidence of a debilitating introspection.

I have not read more than a fraction of the literature discussed in these 'progress' reviews. Nor do I wish to. Even though the radical critique may be overplayed, I also find much of the humanistic writing, especially that focusing on 'sense of place', somewhat precious. No doubt it was a healthy reaction to both the excesses of the positivists and the arrogant social determinism of the radicals, but at least these latter people are dealing with problem issues in the real world, past or present. No one would be justified in suggesting that all contemporary scholarship should be concerned with such issues, but when the discipline of human geography turned to grapple with them, as it did on a large scale after the mid-1960s, it did so with an intellectual arrogance that matches only that of the development economists: two wholly different sets of right answers were offered and insisted upon. Some geographers *have* sought very actively to temper these excesses using behavioural research and argument, but not many such geographers are included among those cited as 'behavioural' geographers in the *Progress in Human Geography* reviews.

Before we move on, it is not without value to consider why anyone should ever have felt it necessary to classify a group of geographers in this adjectival manner. The adjectival classification of geography by subject matter has been with us for a long time; it is also found in other disciplines, though usually to a much lesser degree than with us. It reflects the status of geography as a 'point of view' discipline, bringing its special perspective to a wide area of topic material; and so also does the disconcerting habit of at least a significant minority in our profession to entitle papers grandly '*The* geography of this or that'. In modern times, however, adjectival labelling has acquired a new dimension. Quite apart from subject matter, we have 'quantitative geographers', 'radical or Marxist geographers', 'behavioural geographers', even 'dirty-boots geographers'. Presumably, therefore, we have behavioural/quantitative/ Marxist/whatever - historical/political/economic/urban/industrial/ agrarian/agricultural/whatever geographers. It is little wonder that our discipline – if such it can be called in these circumstances – suffers from an extraordinary lack of cohesion, contempt for the writings of earlier 'unenlightened' generations, and a distressing want of continuity. It is as though, unable to debate, we are forced continually to oppose. Yet there are many geographers who in their empirical research show considerable ability to debate and to synthesize.

What might have been

In a major critique Bunting and Guelke questioned the supposed linkage between the environment-as-perceived and actual behaviour; and in so far as cognitive images are of value in geographical enquiry, they argue, it is as means to an end.[14] Guelke is also a contributor to this volume, and his ideas are further developed in his chapter (see chapter 15). I therefore confine myself to broad agreement with what he and Bunting wrote in 1979, and develop instead the argument outlined above: that the proper core of relevant behavioural study in geography should be in the analysis of decision-making.

Decisions are taken by everyone, but neither all decisions nor all decision-makers are equal; some are much more equal than others. The decision whether or not to build a great dam in the middle of Sarawak, flood out 5,000 Kayan and Kenyah people and resettle them, and generate enough electricity to supply most of peninsular Malaysia by means of the world's longest underwater cable, is clearly of a different order from the decision of an individual to migrate, or to prefer one place rather than another in which to live if he or she had free choice. Decisions such as the former are not taken by one person, and involve a range of clearly measurable considerations as well as a great deal of informed guesswork, or 'weighted judgements' as a colleague of mine has described them. The environment of such decisions as that about the Bakun dam in Sarawak is economic and political: national capacity to bear debt burdens, projections of electricity demand, the desire of politicians to be seen to be doing big things, the structure of politics, the political clout – or lack of it – of Kayan and Kenyah shifting cultivators, the assessments of engineers, the political relations between Sarawak and the Federation; all these and others form part of the decision-making context. Every contributing decision-maker at every level has a 'behavioural environment', but an important part of this environment is economic and structural and the 'information content' of these environments is constantly changing.

The Bakun dam is an extreme illustration, and it seems that the 'new information' of mid-1980s economic recession is likely to kill that project for the time being, but it helps me to make a more general point. All decisions of consequence are the product of intended but imperfect rationality, given the goals and values of the decision-makers. They are also taken in a context of available resources and constraints, and of the conditions which govern the freedom of action of the decision-makers, that is of the political economy – both macro and micro – which surrounds the decision-makers. A large subset of all decisions of consequence have 'downstream' effects: they impact on others. These impacts may or may not have an effect on the decision; they certainly have consequences, which include subsequent decisions which have to be taken by widening chains of people. By the same token, a further large subset of decisions has 'upstream' preconditions, the consequences of former decisions. In this direction the chains that could be traced are almost endless, but once again some past decisions had much greater 'downstream' effect than others.

We noted that, when the behavioural movement in geography began, its perceived role was in planning. In so far as geographers have subsequently involved themselves in risk assessment, in development practice, and in the politicking of environmental issues, among other practical fields, many have been able to bring behavioural considerations to bear on decision. Academic geographers might have done much more than they have done to support these fields of professional application through their research. A 'geographical problem' is clearly enough defined: it concerns organization or distribution in space, or the use of environment, or both. At all levels of resolution from the world economy down to the household and individual, thousands of decisions with consequence in changing spatial organization and affecting environment and environmental process are taken every day. Modern-day geographers have studied several of the structural elements governing such decisions in considerable empirical detail, particularly through the 'business organization' approach[15] and the more widespread but selective fascination with multinationals. Buttimer stressed that the humanistic contribution to the study of decisions should be through the characterization of values, which have profound effects on what is actually implemented.[16] However, studies of particular decisions with geographical consequences have tended to remain in theses and a sprinkling of journal articles. One such, by Juvik, concluded that much more empirical testing of theoretical approaches is needed if theories are more closely to approximate real-world situations; her thesis offers some very striking case material for this contention.[17] However, hers is a plea that one seems to have been hearing for 20 years, and we still await results.

Recently, an attempt has been made by geographers to understand the reasons why land is allowed to become degraded, at general and regional levels of resolution. Blaikie approached the topic with humanized Marxism;[18] Blaikie and Brookfield sought to round this out by infusion of structurally aware behaviourism.[19] It is not for me to pronounce on their success, but they did focus on the conditions of decision-making, and found *both* structural and behavioural approaches to be, together, of value.

Where, in all the literature of the last 20 years, have been the major comparative studies of significant decisions in both spatial and temporal context, exploring the conditions of decision-making, the cognition and values of the deciders, and both the major upstream antecedents and major downstream consequences? There have been some attempts, for example by Gladwin and his associates, though not within geography.[20] The dogmatic arguments of the structuralists have extended over so wide a range of fields that they have thrown up innumerable hypotheses for testing around the counter-hypothesis that the behavioural conditions of decision-making are governed by values and cognition, and by information and the ability to handle it, as well as by the structures which provide the frame.[21] All this has been discussed at high levels of abstraction, but where has been the comparative empirical work? For geographical problems, behaviourally inclined geographers should have been the people to supply it. Had the behavioural movement followed its early promise in this direction we might now be much further forward in generating a strong geographical contribution to understanding the real world of decision and action in the environment. We would certainly still have disputes, but over real matters of interpretation. Certainly more would have come from the behavioural movement than the essentially idiographic 'sense of place'.

Hopeful developments, and a further question

I make no apology for concentrating from this point forward on literature about the non-Western world. I know it better, and it is here that contributions by geographers incorporating behavioural study seem to me to have made most progress. An alternative area of hopeful development might be seen in the new growth of urban ethnography of Western cities since the 1960s, but not much of this work has been done by geographers.[22] There are, admittedly, certain features of the non-Western world which make behavioural considerations in interpretation a necessity rather than an option. The facts of underdevelopment, sharp differences in culture, and a very different political economy make for obvious contrasts between behaviour in these countries and in Western countries, though the contrasts can be overstated. Moreover, interdisciplinarity takes a rather different form in these countries, especially among foreign researchers who meet and compare ideas and findings away from the disciplinary behavioural environment of their home departments, and where research-visa applications have to be justified to authorities who are not interested in intradisciplinary squabbles. In these circumstances not only ideas but also methods rather easily get transferred, and we find geographers or political scientists, even economists, adopting the methods of anthropologists, while the latter in their turn have latterly analysed historical documents and made maps. The influence of anthropology is important, for this is a discipline based on the search for understanding of behaviour and structure, values, and perceptions, which are assumed to differ from those of the researcher. Moreover, anthropologists had a leading role in undoing, or at least undermining, the misconceptions about non-Western people that were held by Western officials, traders, planters, and visitors, and which became a major part of the behavioural environment of colonialism.

With so much to be corrected, the use of cognitive method came easily to those who began to work at micro-level among the rural and urban people of non-Western countries in the 1950s and 1960s. Seeking to establish the environment of behaviour and action, these geographers found it necessary to try to understand social systems, livelihood, and the perception of natural environment and society. They rather eagerly made use of ethno-ecological method once it was introduced by Conklin,[23] and made some forays into the use of ethno-historical method, especially latterly. They were less impressed by the more sophisticated tools imported from psychology by behavioural geographers, and suggestions as to how these might be adapted for use in 'cross-cultural' situations have fallen on rather barren ground.[24] The main methodology was the use of structured observation and enquiry, if not exactly 'participant observation', together with sensitive listening. Added to this was the method of generating hypotheses almost daily to explain this or that aspect of what was observed, small hypotheses that could quickly be tested, and most of which as quickly bit the dust.

I have described and discussed this method, its aims and its possible successes, in more than one place,[25] and I have sought to demonstrate its use in many others. This work from the Pacific is paralleled by other geographical research in many parts of the non-Western world, making greater or lesser use of behavioural approaches, greater or lesser use of structural insights, and with specific emphasis on a wide range of topics studied at micro-level. This is not the place for a comprehensive review of research which includes the work of many continental European and North American geographers as well as Antipodean and British, but it collectively forms a substantial body of results. Very few of its practitioners would describe themselves as 'behavioural geographers'. Most of them would agree with Bunting and Guelke[26] that cognition (and indeed behaviourism as a whole) should fundamentally be treated as a means towards an end, that of better understanding the explicandum of changing human activity on the face of the earth. It may therefore be useful to dwell briefly on the worked examples of one or two geographers of more specifically behavioural intent who have participated in this substantive field of research.

Kirkby [Whyte], a quantitative geographer who later became a leading advocate of cognitive approaches, has made only one major published foray into the field of modern agrarian change, essentially as a by-product of work undertaken with another purpose.[27] Faced with a need to explain fixed goals, or suboptimal production of corn and cash crops in the Oaxaca valley of Mexico, Kirkby employed a variety of decision-making models, and found none which require the assumption of economic rationality to be applicable. She might have gone on to the work of Chayanov,[28] as many of us did, but instead turned to social rationality, in which production decisions are constrained by levelling mechanisms governed by social relations between people, which are valued more highly than individual gain. Not everyone would agree with her in supporting the view that peasant (or primitive) economic behaviour is an aspect of kinship behaviour: Leach, for example, wrote that 'kinship systems have no "reality" at all except in relation to land and property', and elsewhere that 'the kinship system . . . is, at every point in time, adaptive to the changing economic situation'.[29] The interest of Kirkby's work, however, lies not so much in the theoretical strength or weakness of its conclusion, but in the quite rigorous procedure of testing a series of models derived from decision-making theory before reaching a 'social' explanation. Not many have matched her in this.

More recent is an unusual product of the 'natural hazards' school, unusual both in use of ethno-ecological method, and in the social awareness of interpretation. Johnson, Olson, and Manandhar were concerned with the response of Nepalese hill farmers to the very real hazard of erosion and slumping of their elaborately built terraces, both wet and dry.[30] They found that knowledge of the nature of the problem, and of means of repair, were both good, though the supernatural was invoked to explain some sudden events. The larger problem was resources, which most farmers lacked. Thus, they could decide to hang on in the hope that the problem would not become worse while they accumulated resources for complete repair, or cut their losses and abandon the endangered field. It was the distribution of resources for repair in the face of this annual hazard that determined that the net long-term effect of 'random' landslides was to make the poor poorer and the rich relatively better-off.

Watts took an approach through revealed behaviour much further by showing how drought in northern Nigeria impinges very unequally on members of a society made more unequal than it was by the political economy of colonial and post-colonial capitalism.[31] It is the poor who first have to sell their labour, cattle, and assets, then mortgage or sell their land to buy grain. It is the rich who are able to sell grain, buy livestock and labour, and sometimes also buy land. The impact of the hazard is mediated by the socio-economic structures of society; hazard theory – and, by extension, the use of cognitive approaches as a whole – has hitherto attached itself to a specific view of nature, society, and change. Interestingly, all three of these accounts conclude by stressing the effect of 'development' in the increase of risk and isolation, and the decrease of social security. Not many behaviourists, however, as distinct from their critics, have yet gone beyond this to show how structural change and change in values are interrelated, and how both impinge on and differentiate the 'behavioural environment' of individuals. Nor, because of the seeming irrelevance of behavioural methodology to the revealed changes that they study, have many students of agrarian and urban change in non-Western countries, or for that matter in Western countries also, tried to come to grips with the problem of analysing the behavioural environment of persons of different status within the political economy.

A further question is deeply implicit in this brief discussion. It is a fundamental belief of those who support and use cognitive approaches to enquiry and analysis that the perception and behavioural environment of the people whom they study are accurately reconstructed. It is also fundamental to the behaviour-alists that there is a clear link between the perceived image of environment and actual behaviour within it. I earlier supported this view myself. However, if the perceived environment of behaviour is incompletely or inaccurately reconstructed, we not only fail to understand but also have failed to establish any predictive frame for the interpretation – or explanation – of behaviour. Moreover, if the perceived environment, even if accurately reconstructed, excludes important parts of the real environment, then action is likely quickly to run foul of these unperceived or misperceived elements, and a new and probably still-inaccurate 'perceived environment' will be generated. The natural-hazards school alone has focused on this question of misperception, but it has much wider implications. I now turn to these questions in my concluding section.

Whose behavioural environment?

Some lessons from the anthropologists

Leach's disagreement with many of his colleagues over the material role of kinship, mentioned above, is but one of many examples from within anthropology.[32] Leach's assertion was hotly disputed by other anthropologists, though in the context of modern Marxist approaches in that discipline it appears mild and understated. In the broadest sense of the term 'behavioural environment', anthropology is a discipline fundamentally concerned with its delineation and interpretation as a frame for action. Disputes over interpretation of the same basic sets of data are constant, and some collaborating geographers – myself included – have found themselves involved.

Right through the 1960s and early 1970s, for example, there was a lengthy dispute over the role of the principle of agnation in descent-group formation, in relation to behaviour in the matter of land among the highland peoples of Papua New Guinea. It was argued that as land became scarce the principles of agnation and patrilocality were more strictly enforced among societies in which access to land depended on agnatic group membership. Others argued that this was not so, and that interpersonal and intergroup reciprocity continued to demand that migrants could come and go, and transfer their allegiance. The evidence itself was re-examined and recalculated, but also at issue were questions of perception and ideology. It was a tacit assumption that land shortage was accurately perceived, and that this perception influenced measurable behaviour. The ideology of agnation seemed clear enough: men and women are supposed ideally to dwell where they belong in a segmentary descent-based system, and residence elsewhere is a deviation from the norm. What is interesting is that these latter propositions derived not only from observation and enquiry on the spot, but also from the superficial similarity of these societies with those of much of Africa, where structural models of social organization had been strongly developed by mainly-British anthropologists in the preceding decade. Barnes was the first to challenge this derivation,[33] but the argument continued for a further decade as accumulated evidence of observed behaviour slowly eliminated these preconceptions. The evidence remained contradictory, however, and at one point in our involvement as geographers I and one of my students found ourselves on opposite sides.[34]

Since that time these same societies in New Guinea have been re-interpreted with the insights of Marxist analysis.[35] In a good example of eclectic use of theory, the geographers Allen and Crittenden have used these several approaches to develop an elegant hypothesis suggesting that the pressures of a pre-colonial political economy were responsible for the degradation of some of the land.[36] This political economy evolved around the increasingly large-scale use of sweet-potato-fed pigs as indemnifying items in a competitive and status-generating system of exchange, since the sweet potato was introduced some 300 years ago. These pressures, transmitted through farmers under the control of ambitious men, led to unwise use of sensitive land. In all this we are dealing with a *mélange* of observed fact, inferred behavioural environment, and hypothesis illuminated by theory. There is nothing wrong about this, but it is important that we recognize the well-encapsulated place of actual reconstruction of the behavioural environment in the system of explanation. There has been an advance in method, with greater reliance on systematic observation and carefully constructed theory, and less on the inductive case-study method of which Leach earlier wrote:

> Case-history material ... seldom reflects objective description. What commonly happens is that the anthropologist pro-pounds some rather preposterous hypothesis of a very general kind and then puts forward his cases to illustrate the argument. ... Insight comes from the anthropologist's private intuition; the evidence is only put in by way of illustration.[37]

Humanists in geography might do well to consider whether or not they are still in some danger of meriting exactly the same criticism.

Critical comments on the simplest case

The geographers who became involved in the sort of work described above have undoubtedly been influenced by anthro-pological method, and it might be said by their colleagues that the more rigorous use of structured questionnaires, personal con-structs, and tests of cognition might have yielded better results. I doubt this, if only because, as Kirk argued, the behavioural environment is an undifferentiated whole, and the sophisticated tools are able to deal with only a limited part of it. In any case, where we are concerned with multivariate explanation and not merely with description, multivariate enquiry is also necessary.

I implied this much in my earlier review of perception studies, but in that paper I also spent some space on discussion of what should be, for geographers, the simplest case: the reconstruction of the physical part of the perceived environment in terms of the valuation placed on different parts of the total physical resource.[38] My observation that the farmers I studied had a detailed understanding of surface soils and their characteristics has since been replicated so often that it is almost a generalization about farmers everywhere: knowledge of the resources on which they mainly rely is comprehensive, and often closely isomorphic with the most detailed scientific classification. However, I was also seeking perceptions of the value of different types of land, necessary to a hypothesis which would relate the patterns of land-holding and land transfer to this valuation. I reported this exercise in ethno-scientific method in detail and recognized that there were differences in this aspect of environmental perception between different informants.[39] However, I then plumped for one view that reflected both my own perception and the observed land use of the time. I still believe that in terms of a pre-cash economy I was probably right, but the test of time was not available: introduction of coffee with its preference for moist situations, and consequent shifts in the pattern of land use brought out the weighting given to location near where one lives in any perception of environment. In other words, I had failed to take the whole behavioural environment into account.

This is an experience of the sort to give one pause. More recently, working among a far more sophisticated people in Malaysia, but in far less depth, I have become aware of the rapid changes in the total environment of rural people becoming absorbed into the fringes of an expanding city, and of the uncertainties which this has generated. I have not tried any reconstruction of the behavioural environment, but my Malaysian colleague has tested the perception of land as (traditionally) the most valued resource in these contexts, with very mixed results which we have yet to report. It is, however, clear that short-term considerations dominate the perceived environment in such a situation, in which the real long-term behavioural environment contains many elements which are unknown.

The two behavioural environments

Mental constructs are transitory, and sensibly usable by deciders only as a basis for short-term decision. Longer-term actions, whether by an individual or a collective, require much more information and normally efforts are made to obtain more; yet much must remain unknown or imperfectly known, and we are back with the conditions of intended but imperfect rationality. We tend to simplify: 'border-line cases make us nervous and we deal with them by forcing them into one category or another'.[40] No one carries his whole decision-scheme in his or her head. No one is possessed of all the necessary information for rational decision, and no one really takes adequate account of all the possible outcomes.

This being so, as I believe it to be, the behavioural environment of an individual is inevitably imperfect, and not made less so by aggregation with the behavioural environments of others in the same class or situation. Instead of complete cognition there is often a known set of responses, such as the 'hidden transcripts' of minor but persistent resistance to authority, so brilliantly analysed by Scott.[41] It is our task as social scientists not simply to interpret action on the basis of transitory perception, but of such perception in the context of all that we know of environment, political economy, past and forecast trends; it is because of the uncertainties of the latter that it is easier, and intellectually more satisfying, to construct the behavioural environments of past actions, warts and all, from historical evidence and inference.

However we undertake this task it is our own behavioural environment that becomes important. Scott's 'weapons of the weak' derive as much from his own concern with peasant resistance, and his own system of values, as from the words and actions of his informants.[42] Marxists tend to assume a standardized behavioural environment for all members of any social class, together with a system of values: we may disagree with them, but if we undertake research into real cognition to show that they are wrong, we at once and inevitably impose our own system of values. It is not possible to get inside the head of another person, and all evidence suggests that the gifts of a fortune teller or a traditional healer might get us further than any of our scientific methods in trying to do so. This problem is compounded by the fact that even if we fluently speak the same language, or the same second language, as our informants, we are unable to get behind the meaning of every word, every 'yes' or 'no' answer in every box.

The behavioural environment that we need first to study, then, is our own as social scientists. Our own training, value systems, ideology, and preferences impose themselves on any research enquiry. It is important that we recognize this, and that we are projecting our own behavioural environment on to the data. We then need to question the appropriateness of these values and perceptions to the people whom we are studying, and to try to deduce their own. Revealed preferences, through observed actions and their consequences, are the best guide; what people say in answer to questions, or say spontaneously in their conversation, is a valuable supplementary tool, but it needs the solid anchor of observation to be evaluated and used.

Finally, then, I reach a much more negative conclusion about the value of cognition in behavioural research than I did in 1969. The study of cognition may offer insights, but only if we can reach real cognition, and only if we recognize the important role played by the filter of our own perception and system of values. And this is not easy, and perhaps not possible. We are on safer explanatory ground if we give prime place to observation of facts knowingly selected in relation to our stated explicandum, and allow studies of cognition the important but secondary place of generating surprises which then lead us to observe new facts. The best role of cognition is to deepen our use of observation and, by so doing, improve the rigour of the hypothesis-testing which, whether stated or not, underlies all research.

A concluding suggestion may be useful. In his excellent discussion of imperfect rationality, Watkins focused on the use of bungled actions as a test of behavioural theory.[43] Geographers encounter an enormous number of such in their studies of location, spatial organization, and use and misuse of environment. Positivistic and structural theories have been shown, in their use as a guide to behaviour, to have led to many bungles. Perhaps behaviourists among us could concentrate on the study of how they happened? Then perhaps a new adjectival branch of our discipline – 'bungling geography' – could fruitfully emerge.

Notes

1 H. C. Brookfield, 'The geographical study of the "urban sprawl" of western cities', *Indian Geographical Journal*, Silver Jubilee Volume (1952), pp. 101–5.

2 H. C. Brookfield, 'On the environment as perceived', *Progress in Geography*, vol. 1 (1969), pp. 51–80.

3 A. Pred, 'Behaviour and location: foundations for a geographic and dynamic location theory', parts I and II, Lund Studies in Geography, Series B: Human Geography, nos. 27 and 28 (1967, 1969), Gleerup, Lund.

4 D. Harvey, 'Behavioural postulates and the construction of theory in human geography', Seminar Paper Series A no. 6 (1967), Department of Geography, University of Bristol; and D. Harvey, *Explanation in Geography* (Edward Arnold, London; St Martin's, New York, 1969).

5 For example, R. F. Kates, 'Hazard and choice perception in flood plain management', Research Paper no. 78 (1962), Department of Geography, University of Chicago.

6 For example, D. Lowenthal, 'Geography, experience and imagination: towards a geographical epistemology', *Annals of the Association of American Geographers*, vol. 51, no. 3 (1961), pp. 241–60; and D. Lowenthal (ed.) 'Environmental perception and behavior', Research Paper no. 109 (1967), Department of Geography, University of Chicago.

7 J. R. Gold and B. Goodey, 'Behavioural and perceptual geography', *Progress in Human Geography*, vol. 7, no. 4 (1983), pp. 578–86.

8 D. J. Walmsley, 'Positivism and phenomenology in human geography', *Canadian Geographer*, vol. 18, no. 2 (1974), p. 99.

9 For example, E. Relph, 'An inquiry into the relations between phenomenology and geography', *Canadian Geographer*, vol. 14, no. 3 (1970), pp. 193–201; and Yi-Fu Tuan, 'Geography, phenomenology, and the study of human nature', *Canadian Geographer*, vol. 15, no. 3 (1971), pp. 181–92.

10 R. Rieser, 'The territorial illusion and the behavioural sink: critical notes on behavioural geography', *Antipode*, vol. 5, no. 3 (1973), pp. 52–7.

11 E. Waddell, 'The hazards of scientism: a review article', *Human Ecology*, vol. 5, no. 1 (1977), pp. 69–76; and K. Hewitt (ed.), *Interpretations of Calamity from the Viewpoint of Human Ecology* (Allen & Unwin, Boston, 1983).

12 D. Lowenthal, 'Environmental perception: preserving the past', *Progress in Human Geography*, vol. 3, no. 4 (1979), pp. 549–59; T. F. Saarinen and J. L. Sell, 'Environmental perception', *Progress in Human Geography*, vol. 4, no. 4 (1980), pp. 525–48; T. F. Saarinen and J. L. Sell, 'Environmental perception', *Progress in Human Geography*, vol. 5, no. 4 (1981), pp. 525–47; and T. F. Saarinen, J. L. Sell, and E. Husband, 'Environmental perception: international efforts', *Progress in Human Geography*, vol. 6, no. 4 (1982), pp. 515–46.

13 Gold and Goodey, 'Behavioural and perceptual geography'; J. R. Gold and B. Goodey, 'Behavioural and perceptual geography: criticisms and response', *Progress in Human Geography*, vol. 8, no. 4 (1984), pp. 544–50; and B. Goodey and J. R. Gold, 'Behavioural and perceptual geography: from retrospect to prospect', *Progress in Human Geography*, vol. 9, no. 4 (1985), pp. 585–95.

14 T. E. Bunting and L. Guelke, 'Behavioral and perception geography: a critical appraisal', *Annals of the Association of American Geographers*, vol. 69, no. 3 (1979), pp. 448–62.

15 P. McDermott and M. Taylor, *Industrial Organisation and Location* (Cambridge University Press, Cambridge, 1982); and M. Taylor, 'Industrial geography', *Progress in Human Geography*, vol. 9, no. 3 (1985), pp. 434–42.

16 A. Buttimer, 'Values in geography', Commission on College Geography Resource Paper no. 24 (1974), Association of American Geographers, Washington, DC.

17 S. Juvik, 'Environmental conflict and location: the case of the Moomba–Sydney gas pipeline', unpublished Ph.D. thesis, Department of Human Geography, Research School of Pacific Studies, Australian National University, Canberra, 1981.
18 P. M. Blaikie, *The Political Economy of Soil Erosion in Developing Countries* (Longman, London, 1985).
19 P. M. Blaikie and H. C. Brookfield, *Land Degradation and Society* (Methuen, London, 1987).
20 T. N. Gladwin, 'Patterns of environmental conflict over industrial facilities in the United States, 1970–78', *Natural Resources Journal*, vol. 20, no. 2 (1980), pp. 243–74.
21 Pred, 'Behaviour and location'.
22 P. Jackson, 'Urban ethnography', *Progress in Human Geography*, vol. 9, no. 2 (1985), pp. 157–76.
23 H. C. Conklin, 'An ethnoecological approach to shifting agriculture', *Transactions of the New York Academy of Sciences*, vol. 77 (1954), pp. 133–42.
24 A. V. T. Whyte, 'Guidelines for field studies in environmental perception', MBA Technical Notes no. 5 (1977), UNESCO, Paris.
25 Brookfield, 'On the environment as perceived'; and H. C. Brookfield, 'Introduction: explaining or understanding? The study of adaptation and change', in H. C. Brookfield (ed.), *The Pacific in Transition: Geographical Perspectives on Adaptation and Change* (Edward Arnold, London, 1972), pp. 3–23.
26 Bunting and Guelke, 'Behavioral and perception geography'.
27 A. V. T. Kirkby [A. V. T. Whyte], 'The use of land and water resources in the past and present valley of Oaxaca, Mexico', Memoirs of the Museum of Anthropology no. 5 (1973), University of Michigan, Ann Arbor.
28 A. V. Chayanov, *The Theory of Peasant Economy*, ed. B. Thorner, B. Kerblay, and R. E. F. Smith (American Economic Association, Homewood, Ill., 1966).
29 E. R. Leach, *Pul Eliya, a Village in Ceylon: A Study of Land Tenure and Kinship* (Cambridge University Press, Cambridge, 1961), pp. 305–8.
30 K. Johnson, E. A. Olson, and S. Manandhar, 'Environmental knowledge and response to natural hazards in mountainous Nepal', *Mountain Research and Development*, vol. 2, no. 2 (1982), pp. 175–88.
31 M. Watts, 'On the poverty of theory: natural hazards research in context', in K. Hewitt (ed.), *Interpretations of Calamity from the Viewpoint of Human Ecology* (Allen & Unwin, Boston, 1983), pp. 231–62.
32 Leach, *Pul Eliya, a Village in Ceylon*.
33 J. A. Barnes, 'African models in the New Guinea highlands', *Man*, vol. 62 (1962), pp. 5–9.
34 E. Waddell, *The Mound Builders: Agricultural Practices, Environment and Society in the Central Highlands of New Guinea* (University of Washington Press, Seattle, 1972).

35 N. Modjeska, 'Production and inequality: perspectives from central New Guinea', in A. Strathern (ed.), *Inequality in New Guinea Highlands Societies* (Cambridge University Press, Cambridge, 1982), pp. 50–108.
36 B. J. Allen and R. Crittenden, 'Degradation and a pre-capitalist political economy: the case of New Guinea highlands', in Blaikie and Brookfield, *Land Degradation and Society*, pp. 145–56.
37 Leach, *Pul Eliya, a Village in Ceylon*.
38 Brookfield, 'On the environment as perceived'.
39 H. C. Brookfield and P. Brown, *Struggle for Land: Agriculture and Group Territories among the Chimbu of the New Guinea Highlands* (Oxford University Press, Melbourne, 1963), pp. 34–7.
40 J. Watkins, 'Imperfect rationality', in R. Borger and F. Cioffi (eds), *Explanation in the Behavioural Sciences* (Cambridge University Press, Cambridge, 1970), p. 208.
41 J. C. Scott, *Weapons of the Weak: Everyday Forms of Peasant Resistance* (Yale University Press, New Haven, Conn., 1985),
42 Ibid.
43 Watkins, 'Imperfect rationality'.

Index

Re-evaluation

locational studies: behavioural turn
in 207–8; functionalism in 209;
industrial 208–12
Locke, John 278, 279
logical positivism 11
London 107
Lopez, B. 265
Lösch, August 39, 46
Lowenthal, David 72 n99, 236, 249
n5, 296, 311, 312
lunatic asylum 211; in Cumberland
and Westmorland 225–8
Lynch, Kevin 113, 255

McCarthy, J. J. 246
McClung, W. A. 113
Mackinder, Halford 26
McNee, Robert 209
madness: location of institutions
for 205, 225–8
Malaysia 315, 323
Manandhar, S. 319
Mann, M. 246, 247
Mannheim, Karl 51–4, 57, 61, 74
n112
map: as metaphor 257, 261
Marshall, Neil 209
market towns 115–16; post-war
evocations of 118
Marx, Karl 243
Marxism 9, 52, 173, 263, 295, 302,
312, 324; humanized 316
Marxist geography 6, 41, 42, 304,
305
materialism: dialectical 302;
historical 303
Matthiessen, P. 265
meaning 8, 284; humanistic
philosophy of 296; vocational
257
mechanism 256
Megabazus 28, 48
mental maps 12, 15; see also
cognitive maps
Mercator, G. 262
Merleau-Ponty, M. 263, 304
metaphor 15, 256–7; arena as
257; map as 257, 261, 262;

mechanism as 257, 262; organism
as 257; see also sign, symbol
Mexico: Oaxaca Valley 319
Michelangelo 85
Milgram, Stanley 281
millieu 257–8
Milton Keynes 281
mind 4, 5, 11, 16, 27, 38; and
history 297
Minkowski, E. 268
Modernism 175
Moholy-Nagy, Laszlo 170
Moore, E. G. 142
Mounfield, P. R. 207, 208
Mukerjee 20
Mumford, Lewis 85, 163
Murdie, R. A. 262–3
Murdoch, Iris 286
Muscatine, Charles 219

Nairn, Ian 114, 164–7, 170, 175,
176
Nash, Paul 170
nationalism 242, 247
natural hazard 312, 313, 319
Natural Law 19–20, 26
natural regions 12, 40; see also
regions, sea regions
nature 4, 11,16, 27, 42; and
humanity 4–9, 42, 86, 312
naturalism 4–9, 13, 35, 40, 41, 55
neo-Kantianism 57, 61
New Guinea 321–2
New Lanark 106
Newton, Isaac 20, 278
Nigeria 319
Norbert-Schultz, C. 84, 85
nuclear family 245

objectivism 14, 281
objectivity 285; existential 285;
limits to 300
observation 255–6, 325
Olson, C. 129 n1
Olson, E. A. 319
ontology 14
organicism 256; in geography 260–1
organization theory 209

334